HIDDEN AGENDA

NASA and the Secret Space Program

MIKE BARA

Adventures Unlimited Press

**Hidden Agenda:
NASA and the Secret Space Program**

by Mike Bara

Copyright © 2016

ISBN 13: 978-1-939149-66-4

All Rights Reserved

Published by:
Adventures Unlimited Press
One Adventure Place
Kempton, Illinois 60946 USA
auphq@frontiernet.net

www.AdventuresUnlimitedPress.com

Opinions expressed by the author are solely his own and do not reflect the opinions of Adventures Unlimited Press or its officers.

HIDDEN AGENDA

NASA and the Secret Space Program

Adventures Unlimited Press

Acknowledgements

I would like to acknowledge the following beings for their love and support over the years: Dave Bara, Tiffany Masters, Denise Zak, Shari Gaston, Alan Pezutto, Jimmy Church and Rita, and last but certainly not least, Aurora, Miss Fluffy-Muffy, Barkley and Seabass, who make getting up each day a pure joy.

Other books by Mike Bara:
Ancient Aliens and Secret Societies
Ancient Aliens on the Moon
Ancient Aliens on Mars
Ancient Aliens on Mars II
The Choice
Dark Mission (with Richard Hoagland)

TABLE OF CONTENTS

INTRODUCTION   9

CHAPTER 1: FROM PRUSSIA WITH LOVE —
           NYMZA & THE SONORA AIR CLUB   15

CHAPTER 2: NAZI FLYING SAUCERS & THE
           DAWN OF THE FLYING SAUCER   39

CHAPTER 3: MJ-12 & THE NATIONAL SECURITY STATE   65

CHAPTER 4: A NEW FRONT IN A SECRET WAR   87

CHAPTER 5: THE PARALLEL SECRET SPACE PROGRAM   105

CHAPTER 6: THE ALIEN REPRODUCTION VEHICLE   127

CHAPTER 7: THE WHISTLEBLOWERS   145

CHAPTER 8: THE PROJECT SERPO HUMAN-ALIEN
           EXCHANGE PROGRAM   161

CHAPTER 9: ANDROMEDANS, BLUE AVIANS
           AND THE HOLLOW EARTH   179

Dedication

This book is dedicated to Denise Zak, the wonderful, supportive sister I wish I'd had growing up. May God's light always shine upon you. I love you.

See Mike Bara at:

MikeBara.Blogspot.com

HIDDEN AGENDA

NASA and the Secret Space Program

An early drawing from NASA for a proposed Moon Base.

Introduction

"We already have the means to travel among the stars, but these technologies are locked up in black projects, and it would take an act of God to ever get them out to benefit humanity. Anything you can imagine, we already know how to do it."

– Ben Rich, Director, Lockheed Skunkworks

For decades, virtually everyone of significance in the UFO/paranormal field has been using the term "Secret Space Program" to refer to a wide range of mysterious documents, UFO sightings and rumors. While the details of these claims of secret activity vary widely, the consistent factor that they all include is the core belief that the U.S. government, and indeed governments worldwide, have been covering up and continue to cover up the existence of some kind of privately operated space fleet. Beyond that, the term has come to include fanciful stories of time travel, teleportation and alien abduction scenarios that include the cooperation of government agents or secret operatives, like the ubiquitous "Men in Black." The stated purpose of this Secret Space Program is as varied as the rumors, and depending on the claim being made can include defending the Earth against hostile aliens, preparing a great escape for the ultra-wealthy to a new, less troubled home world, or even a grand conspiracy to help aliens mine the human race for our genetic material. Often, these secret plans find their way into coded documents, YouTube conspiracy videos and even Hollywood films. The general tone of these claims is consistently negative; in no case is the program being kept secret for the good of humanity. No, they're out to get us, and even the guys in the government are in on the conspiracy to rob us of our heritage, our lives or even our entire planet.

Back in the 1950s, when rumblings first began through "contactees" like George Adamski and Billy Meier that we had already been to the Moon, Mars and beyond, the tone of the rumors and perspective on the aliens was quite different than it is today. One contactee, George van Tassel, claimed to have met aliens from Venus; he built a giant machine for physical and spiritual healing called the Integratron near Giant Rock, California at their direction. Adamski likewise met with a Venusian named Orthon whom he said communicated telepathically. Orthon informed Adamski that the U.S. government knew all about aliens visiting Earth and making contact at a personal level, and warned him of the dangers of nuclear war. Another contactee, Dr. George King, who later formed the Aetherius Society, also made contact through meditation with an entity from Venus named, you guessed it, Aetherius. Aetherius echoed the warnings given by the other contactees, but his message also included an environmental component, cautioning against the use of nuclear fission (even for peaceful purposes) and arguing that Man was ruining the environment of Mother Earth. Many of King's environmental warnings became staples of the hippie world view of the 1960s. Today however, unlike in the 1950s and 60s, we know that Venus is an uninhabitable hell, and the admonitions and predictions of the environmental lobby have been uniformly proven false. So were the aliens wrong, or did they just reflect the general knowledge and fears of the time?

The first inklings of a Secret Space Program and the conspiracy required to sustain it actually began in 1947. Sightings of "flying disks" accelerated worldwide after World War II, and really took off in the summer of that year when a credible witness, experienced pilot and military veteran, Kenneth Arnold, spotted six disk or crescent-shaped objects travelling in military formation near Mt. Rainier, Washington. His report caused a sensation at the time, but most probably what he actually witnessed was a squadron of back engineered, flying wing fighter-bombers operating out of nearby (and at that time remote) McChord Air Field. At any rate, his report of "flying saucers" made news all over the world, and caused a "flap" of sightings, especially in the U.S. Just a few weeks later, the Army Air Force made a shocking announcement: They had retrieved a crashed flying disk and had moved the debris to a hangar at Roswell Army Air Field near Roswell, New Mexico. The next day, apparently under heavy pressure from Army brass in Washington D.C., the locals retracted their claim and held a press conference in which they displayed the remains of a weather balloon to the assembled press. In 1977, Major Jessie Marcel, who led the team that recovered the debris from a local ranch, revealed that the debris shown to the press that day was not what had been recovered. He said that the debris he and his men picked up was highly advanced and unusual, and he believed it was from another world. He added that he and the other officers at the base, including his boss General Roger Ramey, had been ordered to lie to the press about the saucer and even some alien bodies that had been recovered. Decades later, Ramey's widow confirmed that he had been embarrassed by having to lie to the press about it.

By the 1970s, the Secret Space Program claims had morphed into something far more scary and bizarre, but much more reflective of the times and consistent with the scientific worldview of the day. In June of 1977, at the height of the Watergate scandal and CIA/liberal paranoia, the English TV channel ITV broadcast a faux science documentary entitled *Alternative 3*. It started off investigating the supposed disappearance of notable British scientists, and later claimed to have uncovered a government conspiracy to hide the truth about environmental deterioration and climate change. Things had become so bad, the mockumentary claimed, that the government had been forced to consider three alternatives to dealing with the problem and the inevitable demise of the human race. The first alternative was to depopulate the planet through wars, famine and disease, the second was to build underground shelters for the elite; and the third, known of course as "Alternative 3," was to send the elites to populate the planet Mars using secret alien technology to get there. Alternative 3 was the first reference to the so-called "breakaway civilization" concept in the secret space program mythology, and the concept is still taken seriously to this day.

In the 1990s, the playing field shifted yet again in response to the popularity of TV shows like *The X-Files* and the widespread cultural interest in secret government programs like the B-2 Stealth bomber, which had been developed at ultra-secret bases like Nevada's Area 51. New cultural icons like David Icke emerged into the UFO world, answering all our questions about who was behind all this secrecy and what this amazing technology

Hidden Agenda

was being used for. Depressingly, we were told that hostile races like the Zeta Reticulan "Greys" were harvesting our DNA to create hybrids so their dying race could move in and take over our coveted planet Earth. Behind them, said Icke, were the even more sinister Reptilians from Alpha Draconis who control our ruling elite through secret societies like the *Illuminati*. Some members of the elite, like Queen Elizabeth, actually *are* reptilians themselves.

The Secret Space Program myths and legends are much the same today as they have historically. The breakaway civilization concept is still highly popular with Secret Space Program lecturers and advocates like David Icke, Andrew Basiago and Cory Goode. Just like Van Tassel, Adamski and Meier before them, they have not one shred of proof of their claims. Still, hundreds flock to their lectures to hear fanciful stories of teleporting to Mars to fight wars with the aliens alongside Barack Obama and the like. In Goode's tapestry of whimsical insight, the Nordic Venusians of the 1950s have been replaced by blue Avians from Andromeda who sit at the head table of the Galactic Federation of Light and work in the background to help humanity achieve some kind of spiritual awakening. This awakening is often called "ascension," but it sounds more like assimilation and repression of the individual right to choose which makes us all human. Goode, Basiago and Al Bielek before them also claim that they have been "reverse aged" back some 20 years and had their memories of participating in time and space travel projects like the Philadelphia Experiment wiped clean. However, it appears that the memory wiping technology is not nearly as good as the teleportation or faster-than-light spacecraft, because they somehow broke the psychic bonds and remembered their experiences. Sadly, at least in Goode's case, all this time spent hobnobbing with the Reptilian elites who actually run the world hasn't helped him score much in his return to the real world. Perhaps a book is in the works.

The reader should not take these commentaries as disbelief, but rather simple skepticism of the claims being made by the various witnesses, contactees and whistle blowers. The truth is that there is very little evidence to support many of these specific claims. How, for instance, will any of us ever be able to determine if the Queen is in fact a Reptilian from Alpha Draconis? We won't.

Introduction

But examination of other aspects of the stories—the technologies discussed and explained, the personal testimonies given and a little investigative journalism may uncover facts and papers which show that there may be something more to the story. Something perhaps even a little bit credible. That is what I hope *Hidden Agenda* accomplishes.

The truth is, there is some considerable evidence to suggest that there are or were in fact not one but *two* fully functioning "secret space programs" since perhaps the late 1950s. One of these is the standard and by now almost cliché program being run by the dark cabal of military/intelligence agencies, or what is commonly called the "National Security State." But what I will show with this book is that there is perhaps another thread that developed in response to that program, one that used NASA as a cover to retrieve alien technologies left behind on the Moon and reverse engineer them in an effort to catch up to the bad guys. This program, I believe, was started by John F. Kennedy after he took office in 1961 and included Nazi rocket scientists like Werner von Braun after both he and Kennedy discovered they had been frozen out of the first program by the intelligence agencies. How they responded and the consequences of their actions are questions of great interest to me and hopefully to you, the reader. So let's take it from the top and find the truth behind these rumors together…

Hidden Agenda

Chapter 1
From Russia With Love—NYMZA and the Sonora Air Club

An early "skycycle," circa 1910.

Author's Note: For the purposes of this volume, I will refer to "extraterrestrial" as genetically related to humans or fully human, but not living on planet Earth, as opposed to "alien," which means non-human.

For over a century, the skies over North America and Europe have been filled with reports of unidentified aircraft. They have been referred to as "Airships," "Foo Fighters," "Flying Disks," and now today "UFOs." Speculation has abounded that these craft may in fact have an extraterrestrial if not alien origin. But recent discoveries have shown that many of these unidentified aerial objects may have a much more earthly, if not even more terrifying, genesis.

In the late 1800s there was a wave of sightings of what

Hidden Agenda

First flight of the LZ 1 in 1900.

appeared to be powered, controlled air vehicles in the skies over America. At that time, there were virtually no known flying vehicles at all. Although hot air balloons had been invented a century before, they were extremely rare and generally uncontrollable and had a rounded shape we normally associate with them even today. Widespread use of hydrogen-filled blimps (non-rigid) and zeppelins (rigid airships) was still nearly a decade away, but the reports came in from a vast area of the United States, ranging from Sacramento, California through Texas and even as far east as Chicago and St. Louis. Although Count Ferdinand von Zeppelin had founded the *Gesellschaft zur Förderung der Luftschiffahrt* (Society for the Promotion of Airship Flight) in 1898, the first flight of a functioning zeppelin, the 420 foot long LZ 1 (LZ for *Luftschiff Zeppelin*, or "Zeppelin Airship") wasn't until October of 1900, some five years after the airship flap began.[1]

There are numerous newspaper accounts showing that many of these sightings involved hundreds of witnesses and even landings and interactions with the occupants. Some of these encounters were quite strange, at least as they were reported. According to UFO researcher Jerome Clark, airship sightings were reported worldwide during the 1880s and 1890s, with the first "flap" generally agreed to have occurred in 1887, almost 15 years before the first zeppelin. But the most sensational sightings took place in the United States, mostly in 1896 and 1897. *Wikipedia* summarizes a number of the most well-known reports this way:[2]

From Russia with Love

- The *Sacramento Bee* and the *San Francisco Call* reported the first sighting on November 18, 1896. Witnesses reported a light moving slowly over Sacramento on the evening of November 17 at an estimated 1,000-foot elevation. Some witnesses said they could see a dark shape behind the light. A witness named R.L. Lowery reported that he heard a voice from the craft issuing commands to increase elevation in order to avoid hitting a church steeple. Lowery added that he believed the apparent captain to be referring to the tower of a local brewery, as there were no churches nearby. Lowery further described the craft as being powered by two men exerting themselves on bicycle pedals. Above the pedaling men seemed to be a passenger compartment, which lay under the main body of the dirigible. A light was mounted on the front end of the airship. Some witnesses reported the sound of singing as the craft passed overhead.

Front page of the *San Francisco Call*, November 19, 1896.

Hidden Agenda

- The November 19, 1896, edition of the Stockton, California *Daily Mail* featured one of the earliest accounts of an alleged alien craft sighting. Colonel H.G. Shaw claimed that while driving his buggy through the countryside near Stockton, he came across what appeared to be a landed spacecraft. Shaw described it as having a metallic surface which was completely featureless apart from a rudder, and pointed ends. He estimated a diameter of 25 feet and said the vessel was around 150 feet in total length. Three slender, 7-foot-tall, apparent extraterrestrials (aliens) were said to approach from the craft while "emitting a strange warbling noise." The beings reportedly examined Shaw's buggy and then tried to physically force him to accompany them back to the airship. The aliens were said to give up after realizing they lacked the physical strength to force Shaw onto the ship. They supposedly fled back to their ship, which lifted off the ground and sped out of sight. Shaw believed that the beings were Martians sent to kidnap an earthling for unknowable but potentially nefarious purposes. This has been seen by some as an early attempt at alien abduction; it is apparently the first published account of explicitly extraterrestrial beings attempting to kidnap humans into their spacecraft.

- The mystery light reappeared over Sacramento the evening of November 21. It was also seen over Folsom, San Francisco, Oakland, Sebastopol and several other cities later that same evening and was reportedly viewed by hundreds of witnesses. One witness from Arkansas—allegedly a former state senator Harris—was supposedly told by an airship pilot (during the tensions leading up to the Spanish American War) that the craft was bound for Cuba, to use its "Hotchkiss gun" (a type of gattling gun) to "kill Spaniards."

- In one account from Texas, three men reported an encounter with an airship and with "five peculiarly dressed men" who asserted that they were descendants of the lost tribes of Israel, and had learned English from the 1553 North Pole expedition led by Hugh Willoughby.

- On February 2, 1897, the *Omaha Bee* reported an airship sighting over Hastings, Nebraska, the previous day.

- An article in the *Albion Weekly News* (Nebraska) reported that two witnesses saw an airship crash just inches from where they were standing. The airship suddenly disappeared, with a man standing where the vessel had been. The airship pilot showed the men a small device that supposedly enabled him to shrink the airship small enough to store the vessel in his pocket.

- On April 10, 1897, the *St. Louis Post-Dispatch* published a story reporting that one W.H. Hopkins encountered a grounded airship about 20 feet in length and 8 feet in diameter near the outskirts of Springfield, Missouri. The vehicle was apparently propelled by three large propellers and crewed by a beautiful, nude woman and a bearded man, also nude. Hopkins attempted with some difficulty to communicate with the crew in order to ascertain their origins. Eventually they understood what Hopkins was asking of them and they both pointed to the sky and "uttered something that sounded like the word Mars."

- On April 16, 1897, a story published by the *Table Rock Argus* (Nebraska) claimed that a group of "anonymous but reliable" witnesses had seen an airship sailing overhead. The craft had many passengers. The witnesses claimed that among these passengers was a woman tied to a chair, a woman attending her, and a man with a pistol guarding their apparent prisoner. Before the witnesses thought to contact the authorities, the airship was already gone.

Hidden Agenda

The Aurora, Texas 1897 airship crash.

One of the most famous accounts from the airship wave of the late 1800s comes from Aurora, Texas. On April 19, 1897, the *Dallas Morning News* reported that a few days before, an airship had crashed into a windmill on the property of a Judge Proctor and then fell to the ground. The occupant was described as "dead and mangled" but definitely "not an inhabitant of this world." Like the vessel involved in the Roswell crash more than half a century later, the craft was described as metallic and it had strange hieroglyphic symbols (runes?) inscribed on it. According to the paper, the pilot was given a "Christian burial" in the town cemetery. *Wikipedia* states that "wreckage from the crash site was dumped into a nearby well located under the damaged windmill, while some ended up with the alien body in the grave. Adding to the mystery was the story of Mr. Brawley Oates, who purchased Judge Proctor's property around 1935. Oates cleaned out the debris from the well in order to use it as a water source, but later developed an extremely severe case of arthritis, which he claimed to be the result of contaminated water from the wreckage dumped into the well. As a result, Oates sealed up the well with a concrete slab and placed an outbuilding atop the slab in 1957."

Several critics of the Aurora event have claimed that the event was fabricated in an attempt to draw attention to the town after it had waned due to being bypassed by a new railroad. In a 1979 *Time* magazine interview, a local woman named Etta Pegues

From Russia with Love

claimed the whole story was a hoax (somebody always does) and that no windmill ever existed on the Proctor property. However, an episode of *UFO Hunters* found the base of a windmill over the well, completely discounting her story.

In 1973, MUFON investigators led by local director Bill Case went to Aurora and investigated the claims of "Texas' Roswell." They interviewed two witnesses to the event who were still living, and both insisted the event had really happened and was not a hoax. Mary Evans, who was 15 at the time, told of how her parents went to the crash site and confirmed the discovery of the alien body. Another witness, Charlie Stephens, who was age 10 at the time, told how he saw the airship trailing smoke as it headed north toward Aurora. He wanted to see what happened, but his father made him finish his chores. Later, he told how his father went to town the next day and saw wreckage from the crash.

When Case and the MUFON group examined the crash site, they found a tombstone marker used in the 1897 burial which had a flying saucer shape inscribed on it. Metal detectors indicated a significant amount of metal under the ground, but they were not allowed to excavate and claimed that locals became very hostile when they requested permission to do so. When they returned several years later, ground penetrating radar and metal detectors indicated both the body and the metal were gone.

While disparate and somewhat bizarre, only a couple of

MUFON investigators examine the "alien gravesite" in Aurora, Texas in 1973.

Newspaper article covering the MUFON investigation and the Mary Evans story.

these sightings fall into the category of what might be considered extraterrestrial encounters. Most of these reports appear to be sightings of conventional (although highly advanced for the time) experimental aircraft. The problem is that although these "airships" were stretching the limits of the known technology of the day, they were not beyond the reach of some well-to-do genius who decided he wanted to conquer the skies. But who at that time might have fit such a description?

Suspicion during the west-to-east airship wave of the late 19th century naturally centered on inventor and industrial mogul Thomas Edison. Edison was one of the few men in the United States at the time that possessed the financial wherewithal and intellect to design, build and test such revolutionary technologies. He was also the one man who might seek to shock the world by holding back before announcing he was behind the whole thing. But he didn't.

Edison only added to the mystery when he took to the pages of the *New York Times* to strongly deny he had any involvement with the sightings that had taken place. In fact, he expressed the opinion that he didn't consider the idea of airships to be commercially viable, and insisted he would have nothing to do with them.

So if wasn't Edison or some other genius inventor, like

From Russia with Love

Thomas Edison's denial of involvement with the 1896-1897 airship flap.

Tesla (who had no money), exactly who might have been behind the sightings of the mysterious airships of the 1800s? In fact, the public's suspicions may have been correct, even if their ideas as to the minds behind these "genius creations" were woefully naive. Some researchers, notably Olav Phillips, believe that these sightings were actually linked to an airship initiative launched more than half a century earlier by a mysterious consortium of Prussian nobility.

Records show that in the 1830s, there was a private project launched in Prussia called the NJMZA or NYMZA. It was funded by a number of mysterious and wealthy German benefactors, like Hugo Henckel von Donnersmarck and Guido Henckel von Donnersmarck of Silesia, with the idea of developing and exploiting airship technology in the Americas. They soon financed and recruited a number of Prussian/German theoretical physicists of the period, including the likes of Wilhelm Weber and Rudolf Kohlrausch.

There are many theories as to the origin of the name "NYMZA," with some of them falling significantly into the esoteric. Some claim it is simply an acronym, which is common in aerospace engineering and that it stands for some German/English variation of "National Exploratory Airship Program Office," or "New York Mechanical Zephyr Association." Others say the "Z" could refer to Zeppelin, as in Ferdinand von Zeppelin, the inventor

Hidden Agenda

of the rigid airship, and that the "A" could mean "Aero" (a word you'll come to understand shortly) or "Airship." So it could be perhaps the New York Mechanical Zeppelin Association, or even the New York Mechanical Zeppelin Airship. Others think it has a much more ancient and perhaps sinister meaning.

According to some sources, "NYMZA/NJMZA" is a word from very ancient times that predated all known languages but had survived into ancient Egyptian, Greek and Latin in altered forms. It translates loosely to "The Nameless Ones" in today's languages.[3] Associated with evil extraterrestrial "gods" the Sumerians called the Anunnaki (see my book *Ancient Aliens and Secret Societies*), the NYMZA lost a war on Earth for control of the planet to others of their kind and were banished to a kind of netherworld of non-physical existence, something like the "Phantom Zone" in the Superman mythology of Krypton. From this astral plane of existence the NYMZA cannot escape unless humans on the physical planes invoke their true names, which are unknown. However, it is believed by some esoteric schools of thought that the NYMZA can communicate with corporeal human beings living on this physical plane through dreams or séances. According to some Theosophical documents from the early 19th century, séances were conducted to contact the NYMZA, sponsored by the likes of the von Donnersmarck family in an effort to gain scientific knowledge from the exchange. This led to these wealthy industrialists looking at many ancient philosophical and religious texts through the eyes of physicists and engineers. By studying ancient Vedic and Hindu texts like the *Samarangana Sutradhara* (written by Paramara King Bhoja of Dhar, 1000–1055 AD), they began to understand the basic theories behind aircraft called *Vimanas*, which were described in the ancient Vedic sagas. The Indologist George William Frederick Villiers, 4th Earl of Clarendon, actually translated the *Samarangana Sutradhara* into English and provided a description of a "mercury vortex engine" which powered the Vimanas:

> Inside the circular airframe, place the mercury-engine with its solar mercury boiler at the aircraft center. By means of the power latent in the heated mercury which

From Russia with Love

sets the driving whirlwind in motion, a man sitting inside may travel a great distance in a most marvelous manner.

Four strong mercury containers must be built into the interior structure. When these have been heated by fire through solar or other sources, the Vimana (aircraft) develops thunder-power through the mercury. It is also added that this success of an Indian scientist was not liked by the imperial rulers.

What Clarendon was doing was reciting the ancient texts as they were written, in a matter-of-fact manner. It is simply astonishing that the ancient Vedas spoke so directly about flying machines and their power units which, according to current belief, simply could not have existed in ancient times.

Another such manuscript, the *Vaimānika* Śāstra, emerged in the 1920s and was said to have been accumulated from many far more ancient manuscripts authored by Hindu/Vedic *rishis* thousands of years before. As described in news reports from Rueters and other news services of the period:

Mr. G. R. Josyer, Director of the International Academy of Sanskrit Research in Mysore, in the course of an interview recently, showed some very ancient manuscripts which the Academy had collected. He claimed that the manuscripts were several thousands of years old, compiled by ancient rishis, Bharadwaja, Narada and others, dealing, not with the mysticism of ancient Hindu philosophy of Atman or Brahman, but with more mundane things vital for the existence of man and progress of nations both in times of peace and war.

One manuscript dealt with Aeronautics, construction of various types of aircraft for civil aviation and for warfare. He showed me plans prepared according to directions contained in the manuscript on Aeronautics of three types of aircraft, or Vimana's, namely, Rukma, Sundara and Shakuna Vimanas. Five hundred slokas or stanzas dealing with these go into such intricate details about choice and preparation of metals that would be suitable for various

Hidden Agenda

parts of Vimana's of different types, constructional details, dimensions, designs and weight they could carry, and purposes they could be used for.

Mr. Josyer showed some types of designs and drawing of a helicopter-type cargo-loading plane, specially meant for carrying combustibles and ammunition, passenger aircraft carrying 400 to 500 persons, double and treble-decked aircraft. Each of these types had been fully described.

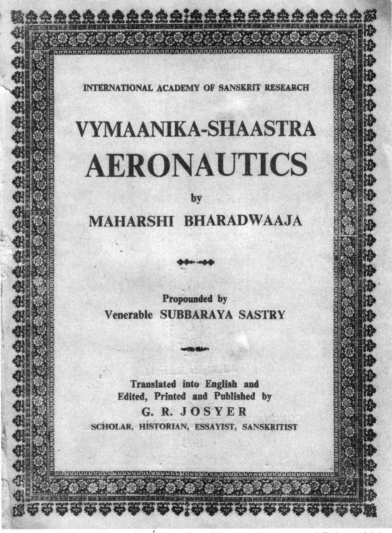

Title page from the *Vaimānika Śāstra*, describing ancient powered flying vehicles.

In the section giving about preparation and choice of metals and other materials that should go into such construction of aircraft, details were specified that the aircraft, (these metals are of 16 different alloys), must be "unbreakable, which cannot be cut through, which would not catch fire, and cannot be destroyed by accidents." Details as to how to make these Vimana's in flight invisible through smoke screens are given in Vimanasastra of Maharshi Bharadwaja.

Further description and method of manufacturing aircraft, which will enable pilots not only to spot enemy aircraft, but also to hear what enemy pilots in their planes were speaking, on principles akin to radar, have all been given in elaborate detail with suitable explanatory notes. There are eight chapters in this book which deal with construction of aircraft, which fly in air, go under water, or float on water.

Obviously, these types of ancient texts, containing detailed descriptions of materials, technology and "mercury vortex engines" might have been of intense interest to a group like NYMZA, which was seeking to unravel the secrets of functioning aerial craft at the very least. These studies ultimately led to early exotic physics experimentation in an effort to replicate the air vehicles of the ancients. Like the German empire and the later Nazi regime, the Kingdom of Prussia was militaristic and expansionist minded, and domination of the air would certainly have been a major boon to the cause of Prussian dominion over Europe. But this effort apparently led them to begin doing research not in their native lands, but far, far away in the Americas.

In the 1850s, NYMZA sent a man named Charles August Albert Dellschau from Prussia to the U.S. to begin scouting sites and recruiting new members to advance their agenda. Dellschau had been born in 1830 in Berlin, but the first definite documentation of Dellschau's presence in America is from 1860, when he applied for citizenship from his home in Fort Bend County, Texas. That form makes reference to an earlier "declaration of intent" from

Charles August Albert Dellschau.

1850, indicating that this is when he arrived in America.[4] The question is what exactly he was doing and where he was for the intervening decade.

Unlike Weber and Kohlrausch, Dellschau was an engineer, a mechanically inclined man who was capable of transforming theoretical concepts into practical, working inventions. Upon his arrival in the U.S., around the 1849-53 period, Dellschau traveled first to Texas and then later to California, where he was instrumental in the founding of the mysterious Sonora Air Club, an offshoot organization of NYMZA. It seems that Dellschau had been given some of NYMZA's theoretical research and was sent to America probably to do actual engineering of the discoveries—or the glimmerings—of the parent group. Under this speculative scenario, the SAC was the engineering and manufacturing arm of NYMZA.

Not a great deal is known about the Sonora Air Club, except that it was founded shortly after Dellschau arrived in the Sacramento/Sonora area, and that its known members were mainly of German/Prussian descent. Records exist of several members of the Club that Dellschau references in his surviving papers. One was a sheriff by the name of James Steward and another an innkeeper named Freund, who historians confirm as

From Russia with Love

"well-documented." There are also references to a Peter Mennis, who served in the Texas Mounted Volunteers during the Mexican-American War, died on November 1, 1901 and is buried in Napa. Mennis is mentioned by Dellschau as the SAC member who invented a "lifting fluid" which made air travel possible. Another member referenced by Dellschau is a fellow Prussian named Gustav Freyer, but few records of his time with the Sonora Air Club have been found. All that is known of him, besides the references in Dellschau's papers, is that he arrived from Germany/Prussia in 1859 at the age of 20.[5]

By most accounts, Dellschau, after his 10 years in California, settled in Houston, Texas where he made his living as a butcher. In 1861 he married Antonia Hilt in Richmond, Texas and had three children and one stepdaughter with her. But, unbeknownst to his friends and family, he was also busy working on a major project involving "Aeros," or airships, right up until his death in 1923.

After his retirement in 1899, he filled at least 13 notebooks with drawings, press clippings, watercolor paintings and collages depicting fantastical airships, which he said were known by the

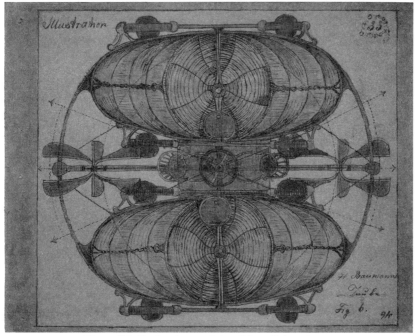

One of Dellschau's surviving "Aero" drawings.

Hidden Agenda

SAC as "Aeros." Dellschau's earliest known work is a diary dated 1899, and the latest is an 80-page book dated 1921-1922.

According to *Wikipedia*: "His work was in large part a record of the activities of the Sonora Aero Club, of which he was a purported member." Dellschau's diary makes reference to one of the members of the SAC (allegedly Freyer) making a breakthrough discovery of the formula for an anti-gravity fuel he called "N/B Gas."

Having cracked this formula, Dellschau asserts that the rest of the energy of the Sonora Air Club was directed to designing and building the first navigable aircraft using the N/B Gas for lift and propulsion. Having apparently achieved this aim, Dellschau moved to Texas and as we have already learned, married. In the last part of his life he made the technical drawings of many NYMZA/SAC airship designs, many of which bear a striking resemblance to newspaper descriptions of the airships in the 1896-97 airship sightings wave in the western United States.

Dellschau's papers and drawings were lost after his death for nearly a half century until they were discovered in the embers of a house fire in Houston in the late 1960s. Dumped on a curb,

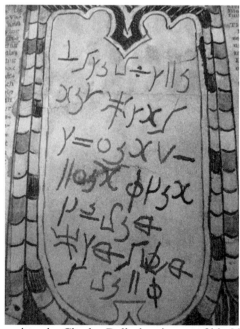

Coded texts written by Charles Dellschau in one of his 80 notebooks.

From Russia with Love

Germanic/Norse "Runic Alphabet" used in interwar Nazi mysticism.

the notebooks were purchased by a junk dealer and eventually purchased by a collector named Pete Navarro, who studied them intently for more than 15 years and partially deciphered some of Dellschau's coded text.

What's interesting to me about Dellshau's "coded texts" is that they bear a striking resemblance to Nazi/Theosophical runic symbols, which were thought by the Vril Society and later Nazi mystics to embue physical objects with magical, physical-law defying characteristics, including a powerful levitation or anti-gravity effect.

Not much is known about "Vril" or the Vril Society, but it is said to have had a major influence on Hitler, Himmler and others within the Nazi elite. The word was first coined in an 1871 novel by Edward Bulwer-Lytton (the first author to coin the phrase "it was a dark and stormy night…") entitled *The Coming Race*. The

Hidden Agenda

novel was later retitled *Vril, the Power of the Coming Race*, and contained many themes that would find their way deep into Nazi mysticism. Vril is defined in the novel as an energy force used by angelic subterranean beings that call themselves the Vril-ya, and intend to someday emerge from their subterranean lair and occupy the surface of the Earth by whatever means necessary. These themes were seized upon by Theosophists like Helena Blavatsky, and later the Thule society, both of which were hugely influential in interwar Germany. According to their versions of the story of this race of subterranean "gods," the Vril-ya were descendants of the survivors of Atlantis who had an Egyptian styled civilization and used to roam the Earth.

The runic alphabet dates from at least the 2nd century AD, and is thought to have been used more as a spellcasting or magical language than for everyday communication. There are Germanic, Anglo-Saxon and Nordic versions of the runic alphabet, some of which are still in use today on charm bracelets or New Age readings.

As far as anyone can tell, the Vril Society used magic spells, runic symbols and incantations to conduct séances and other magical workings. Surfacing in the 1920s (although some scholars dispute whether it actually existed at all) the Vril Society was evidently led by a high priestess named Maria Orsic, an

The only confirmed photograph of Maria Orsic.

From Russia with Love

exceptionally beautiful young woman of pure Aryan blood. Orsic was born in Vienna in 1895 and became well known in postwar Germany as a powerful psychic medium.

She formed the *Vril Gesellschaft*, or Vril Society, in 1921 as a means to study the magical energies of the Vril as outlined in Bulwer-Lytton's novel. She believed, as had been suggested in the novel, that women were the more powerful conduits of Vril energy, and that female hair acted like a kind of psychic antenna amplifying the Earth's magnetic field and placing women in closer touch with the source energies of the Vril. For the reason, Orsic and the other high priestesses of the Society grew their hair as long as possible in order to more effectively channel and utilize this psychic energy.

The Vril Society was also said to be in communication with Aleister Crowley's Hermetic Order of the Golden Dawn, and learned some tips about the spiritual power of "sex-magic" from them. This led Orsic and the Vril Society to conduct orgies and séances—sometimes both at the same time, in an effort to

Photograph believed to be Maria Orsic, showing her amazingly long hair, from a Vril Society handbook.

Hidden Agenda

Photograph believed to be Maria Orsic, assuming a sexually submissive pose in order to draw in mystical energies.

communicate with otherworldly beings.

During one of these affairs, Orsic went into a trance and began to recite passages in a strange language. Later, she wrote down several pages of script in what was eventually determined to be a mish-mash of ancient Sumerian and a semi-secret Templar script, neither of which she was fluent in. After some study, it was determined that she had written down detailed instructions for the creation of a flying ship, a disk-shaped vehicle which the Society came to believe was a plan for a spacecraft from a race of Aryan beings from Aldebaran, a star in the constellation of Taurus the Bull. It is these notes that bear a striking resemblance to Dellschau's "coded texts." While Dellschau makes no mention of the SAC's primary goal being the construction of a spacecraft

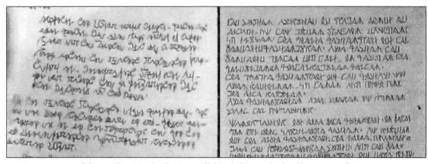

Two pages of the script Maria Orsic scribbled down during her most famous trance.

From Russia with Love

for a trip to Aldebaran, his codes remain largely unbroken and it is far from clear what they contain.

Many skeptics deride Dellschau's work as the overactive imagination of an aging man, who fantasized tall tales of the mythical Air Club and its members. They prefer to refer to him as an "artist" rather than what he claimed to be, which was the draftsman for the Sonora Air Club. Obviously, there is quite a difference between a draftsman and an artist, one recording only what he has seen and one imagining that which does not exist. Dellschau's critics, who simply don't want to believe his story,

Variations on an Aero design concept by Charles Dellschau.

Hidden Agenda

try to characterize him as the latter, but as a former aerospace engineer I can tell you that his drawings are indicative of a broad technical knowledge and understanding of basic design concepts. His drawings are what we once called "production illustrations" in the aerospace industry, depictions of actual full scale designs for use as a guide to then create detailed drawings.

That, and the fact that he used coded texts to protect certain information on the drawings indicates, at least to this author, that he was depicting real experiments and craft from his period working with the Sonora Air Club. This is exactly the conduct I would expect of a member of a secret esoteric society, like NYMZA, and hardly the behavior of a delusional old man fantasizing about a past in California that never existed.

But where did these ideas for anti-gravity propulsion and "Aeros" come from? I speculate that the various members of NYMZA, German industrialists and mystics, got the ideas from ancient texts and then sent engineers—like Dellschau—to the U.S. where they could work in secrecy and freedom.

At first, the idea of a group of Prussian engineers hanging out in a gold rush boomtown 130 miles east of San Francisco to develop secret aircraft might seem unlikely, but in reality, at the time it was the perfect place for such research to have been done. Even today, the American West is largely uninhabited, which is why it is the perfect location for secret bases and military test facilities like Area 51. Given the constant state of war in Europe at that time and the amount of freedom from government interference, wide open spaces, an industrial base and respect for privacy that was available in the U.S., it makes total sense for a secret society—which is what NYMZA is—to want to set up shop there.

People today don't appreciate that the separation of the scientist and the mystic is a very recent phenomenon. In the 1800s and in fact all throughout human history, they were one and the same. It's a very modern conceit that guys like Deepak Chopra and Carl Sagan are in philosophical competition with each other. Back then all the best scientists, like Maxwell and Tesla, were also mystics. So the notion that NYMZA was a collection of "mystical scientists" trying to crack the problem of controlled flight is one I'm very comfortable with. Given that, I'm going to assert the

From Russia with Love

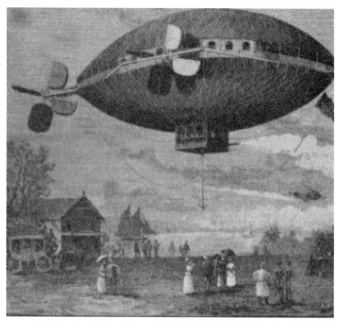
Airship depiction from the 1896-97 wave of sightings.

following speculative scenario to explain the airship wave of 1896-97.

I suspect that after the "N/B Gas" breakthrough around 1860, when Dellschau returned to Texas, the SAC probably went into hiding for a bit. The American Civil War and subsequent westward expansion of the United States probably created a lot of chaos, and it wasn't until the 1880s that things settled down to the point that California was a comfortable place to resume experimentation. By the 1890s, the SAC had developed working airships, and that's when the wave of sightings began to occur.

What's fascinating about the airship wave of the late 1890s is that it originates in Sacramento, very near where Dellschau's Sonora Air Club was presumably conducting their experiments, and then it seems to track across the western U.S. into Texas and places like Aurora, and then on east as far as Chicago. This is the exact path that Dellschau took from Prussia on his way to California. It's almost as if the airships were backtracking along his original route. Maybe they were taking the airship technology they had developed back to Prussia?

After these early experiments with powered airships and the mysterious "N/B Gas" propellant, the mystery airships seemed

Hidden Agenda

to disappear for several decades, until the rise of Nazi Germany. Even during the interwar period when the many esoteric societies emerged in Germany, there was very little in the way of unsolved aircraft sightings. But once the war started, all that changed very dramatically.

(Endnotes)
1 "Die Ausführung des Zeppelin'schen Luftschiffes" (in German). *Die Welt* (Vienna), Issue 22, 3 June 1898, p. 6. Retrieved: 11 March 2012.
2 https://en.wikipedia.org/wiki/Mystery_airship
3 *Empire of the Wheel III: The Nameless Ones*, Walter Bosley
4 http://www.theatlantic.com/technology/archive/2013/03/charles-a-a-dellschau-dreams-of-flying-the-amazing-story-of-an-airship-club-that-might-never-have-existed/274170/
5 http://www.germanimmigrants1850s.com/index.php?id=61679

Chapter 2
Nazi Flying Saucers and the Dawn of the Flying Saucer

A mystery airship from the 1920s.

After these early experiments with powered airships and the mysterious "N/B Gas" propellant, the mystery airships seemed to disappear for several decades. We really don't see much more about airships again until the interwar period, when propeller airplanes, dirigibles and zeppelins became more common.

There were some reported sightings of cigar shaped craft in the 1920s, but these could have been sophisticated zeppelins, and these sightings quieted down around the time Einstein started to work on equations for the Metric Tensor Torsion field (also called the Einstein-Cartan Theory of Gravitation) in 1928. In short, the torsion field theory attempted to unify all of the existing four forces of nature (gravity, electro-magnetism, and the strong and weak nuclear forces) under one umbrella: spin energy, or "torsion." It proposed that spin was the fundamental underlying energetic force in the Universe, and that all other reactions, forces and matter were influenced, if not controlled, by it. See my second book *The Choice* for more information.

After Einstein left Germany in 1933, it is widely believed

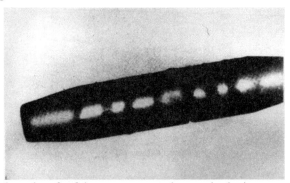

Cigar shaped craft of the type commonly seen in the interwar period in Europe, South America and the United States.

that his colleague Erich Schumann took over his work. While Einstein fled Hitler's Germany, Schumann enthusiastically joined the Nazi party in 1933. Schumann quickly became "extraordinarius professor of experimental and theoretical physics" at the University of Berlin where he not only was awarded a military rank, he was placed in charge of Germany's project to develop the atomic bomb. Schumann was also academic advisor to a brilliant young engineer named Wernher von Braun, and helped him with his doctoral thesis.

While much of the Nazi war machine was focused on developing tanks, guns and even the atom bomb, a separate organization, perhaps influenced by the earlier NYMZA research, was created under the command of SS General Hans Kammler. Eventually rising to the rank of SS Ubergruppenfurher, Kammler was in charge of all the most secret Nazi weapons programs, including the ME-262 jet fighter, the V2 rocket, and perhaps the most mysterious of the super-secret Nazi weapons, "Die Glocke," otherwise known as the Nazi Bell.

Of all the Nazi superweapons of World War II, the most intriguing has to be the so-called "Nazi Bell." It was a high-level experimental device designed to exploit very exotic physics theories involving counter-rotating magnetic fields. It appears to be a generator or reactor of sorts that could be used to create anti-gravity craft, like flying saucers, or perhaps horrifically powerful beam weapons, like amped-up lasers.

It is now widely believed that the Nazis were developing these exotic weapons in an area of Prussia called Silesia, where

SS General Hans Kammler.

a mysterious "rubber plant" that was using vast amounts of Germany's energy resources was located.[1] For reasons that make no sense strategically, Hitler insisted on protecting this area even over the German capital of Berlin in the closing days of the war. The story is that the Bell was being developed by Third Reich scientists working for the SS in a specialized German facility known as *Der Riese* ("The Giant"). Today, all that remains of the Die Glocke is a concrete test rig nicknamed "The Henge" that lies abandoned.

Obviously, no matter how mad with syphilis he'd become, Hitler wouldn't have committed those kinds of military resources to protect a rubber plant. There was clearly something important going on there, and Hitler kept insisting that the research had to be protected because there were superweapons being developed there would save the war effort for the Germans.[2] Even an atomic bomb or two mounted on a V2 rocket wouldn't have ended the war in Germany's favor. He had to be talking about something else, something like the Bell.

Stories about the Bell's effects range from the disturbing to the horrific. It is said to have killed anyone who came in contact with it, and it is said to have been able to kill anyone and anything for hundreds of meters around the test rig. There are stories from Igor Witkowski, a Polish writer who read East German transcripts of the debriefing of Nazi scientists about the Bell and its effects. Witkowski says Die Glocke, when activated, had an effect zone extending out as much as 660 feet. Within the zone, "crystals would form in animal tissue, blood would gel and separate while plants

Hidden Agenda

Depiction of the Nazi Bell inside The Henge, a concrete test rig constructed to contain it.

would decompose into a grease like substance." [3] Witkowski also said that five of the seven original scientists working on the project died in the course of the tests.

It was also said to have strange runic hieroglyphs engraved on its base, and it wasn't until these were added along with a ceramic shell that it became safe to work around the Bell without your blood being boiled inside your body.

How exactly it worked is something of a mystery, but it is thought to have contained massive counter-rotating disks or plates that, when mixed with a mysterious, thick violet fluid called "Red Mercury" or "Xerum-525," induced an intense electromagnetic field that somehow cancelled out both gravity and inertia. According to Witkowski, Die Glocke is described as being a device "made out of a hard, heavy metal" approximately 9 feet wide and 12 to 15 feet high, and having a shape similar to that of a large bell. The metallic liquid code-named Xerum-525 was supposedly "stored in a tall thin thermos flask a meter high encased in lead." Additional substances said to be employed in the experiments were referred to as *Leichtmetall* (light metal) and included thorium and beryllium. The effect it generated supposedly gave it the capability to defy gravity and fly at speeds that defied the laws of physics. Armed with even conventional guns or bombs, it would have been invulnerable to Allied attacks.

What's interesting about this Xerum-525 or Red Mercury as it is known today, is that it sounds an awful lot like Dellschau's "N/B Gas" which was supposed to have given the early airships anti-gravity capability long before the invention of flight. The massive counter rotating plates or cylinders, perhaps made of the exotic thorium and beryllium, are right out of the Einstein-Cartan torsional playbook. The Bell seems to be nothing but a highly advanced anti-gravity reactor based on technologies derived from the N/B Gas experiments and Einstein's speculative and nearly forgotten physics equations.

So we have this path from Prussia to California in the 1800s, the discovery of N/B Gas, and then this west to east pattern of flying ships, as if the secrets were being flown back to Prussia. Then we see the re-emergence of N/B Gas in the form of Xerum-525 or Red Mercury in the Nazi Bell, again in Prussia, about 40 years later.

Could the Red Mercury that powered the Nazi Bell be a further development of the "N/B Gas" discovered by the NYMZA scientists and developed over a 40-year period? And if so, what did the Bell experimentation lead to?

One thing is certain: Whatever was going on at Der Riese

Artist's depiction of the Nazi Bell.

Hidden Agenda

General George S. Patton, commander of the U.S. 3rd Army

was of sufficient concern to the Allied High Command that they sent their best army directly to The Giant in a desperate effort to put a stop to what was going on there.

General George S. Patton was the United States' best military commander in World War II. A gruff, aggressive soldier, Patton had fallen out of favor with the Allied High Command and General (later president) Dwight Eisenhower because of his flamboyant style and his tendency to say things that were politically incorrect. In virtual exile after several incidents that made the press angry, Patton was recalled in 1944 when the Allied forces found themselves in a desperate situation.

Trapped in the hedgerows of Normandy, Patton's commander General Omar Bradley had devised a plan called

Nazi Flying Saucers & the Dawn of the Flying Saucer

Operation Cobra to break out and liberate France. While most of the forces concentrated and bombarded the center of the German lines, the U.S. 3rd Army, led by Patton, swung around in an end run on the German left flank. Patton's army devastated the German defenses, broke out and raced across France all the way to the Rhine River several months later. In December 1944, the Germans launched a massive counter attack (the "Battle of the Bulge") against a weak point in the Allied lines and threatened to drive to the sea and cut the American and British forces off from each other. Patton, engaged in heavy fighting south of the Bulge, ordered six divisions of his 3rd Army to disengage and pivot 90 degrees north to attack the German left flank and relieve the besieged 101st Airborne Division encircled at Bastogne. Eisenhower and Bradly, believing such a maneuver to be impossible, held Patton back for several days before finally letting him engage. Patton's pivot and drive with over 133,000 men is widely considered one of the most extraordinary military maneuvers of all time.

After the Battle of the Bulge, the Germans retreated to their borders and fortified their defenses on the Siegfried Line behind the Rhine. On February 23, 1945, Patton's forces breached the Siegfried Line between Remagen and Frankfurt. At this point,

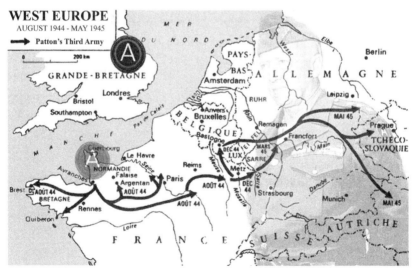

Map of Patton's 3rd Army's advance across Europe showing his diversion to Silesia. Patton could have easily beaten the Soviets to Berlin, the biggest political and economic prize of World War II.

Hidden Agenda

the road to the major political prize of Berlin and the subsequent decisive defeat of Nazi Germany was wide open. Patton, by his own estimates, could have carved through the scattered German defenses and sacked Berlin in two weeks. But instead of bringing supplies forward to support his drive on Berlin, Eisenhower and Bradley ordered Patton to stop right where he was. Instead, they diverted his armies southeast...

Toward The Giant, and Die Glocke.

In political and military terms, this makes no sense. Berlin was *the* political and economic prize of World War II, and had the Allies captured it before Stalin they would have been in a much better position to dictate postwar terms to the aggressive, expansionist Soviet dictator. Instead, Patton was first stopped and then diverted to Silesia, driving right toward The Giant and that mysterious rubber plant. Patton could have easily taken Berlin well ahead of the bogged down Soviet armies. But obviously Allied High Command considered that Silesia, the area Hitler was inexplicably protecting, was more important than Berlin.

It is of course speculation to conclude that Eisenhower was after the Nazi Bell or other secret weapons, but what else of value was in the area? A few concentration camps and a rubber plant? To my mind, the diversion of Patton's forces is the strongest possible argument that there was indeed something very valuable that the Allied forces needed to get their hands on.

Of course, later in the war especially, you begin to get reports of the so-called "Foo Fighters" over Germany, UFOs that exhibited all of the performance characteristics of modern-day

A rare photo of "Foo Fighters."

flying saucers. There were clusters of them over Germany and some said they flew in formation with Allied aircraft, but never displayed hostile intent. It seems possible that the Foo Fighters were somehow connected to the experiments taking place at The Giant, but they remain a mystery today.

At the same time, another program was being developed under Kammler's command by another German inventor, Viktor Schauberger. He was working on a device he called the Repulsine, which used a vortex generator to exploit the Coanda Effect into lift. He actually designed several Repulsine devices which achieved flight using these principles. His grandson even has a working model of one the Repulsine devices today.

Is it possible that Kammler's secret organization found a way to exploit Schauberger's ideas into a highly advanced, working Nazi super weapon?

The problem with the Repulsine was that, while the idea was sound, there was no conventional way to generate enough power to really make it work. Now, if you had a sort of hyper-dimensional torsion field reactor, like say the Bell, as the power source, you could combine that with a Repulsine air frame and you might be onto something. When you combine these three

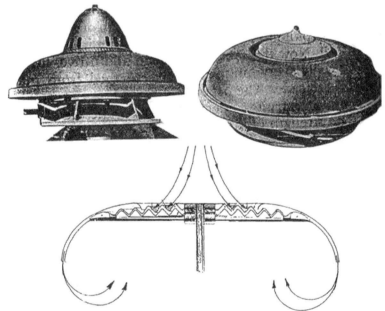

Viktor Schauberger's "Repulsine" design.

Hidden Agenda

ideas—the N/B Gas or Red Mercury fuel, the Nazi Bell counter rotating magnetic fields which cancel inertia and gravity, and the disk-shaped Repulsine airframe, which allows for directional control—you have all the ingredients you need for the proverbial

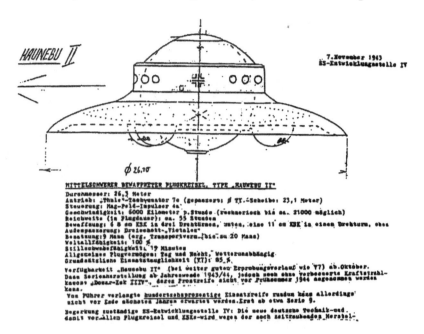

Nazi Flying Saucer!

After the war ended, Kammler and all references to the Bell simply disappeared. Officially, Kammler was reported dead four different ways, and at four different times, so it's safe to say nobody knows what happened to him. There are strong rumors that he fled to South America along with the Bell prototype and from there moved on to a secret Nazi base in Neuschwabenland in the Antarctic.

Indeed, only months after the war in Europe had supposedly ended, the United States sent a task force of over 4,700 men in 13 ships and 33 aircraft to Antarctica on a mission called Operation Highjump under the command of renowned arctic explorer Rear Admiral Richard E. Byrd. The operation amounted to nothing less than an invasion of Antarctica. The task force included one aircraft carrier, two destroyers, two seaplane tenders and a variety of support ships, including two ice breakers. Upon arriving on the

Nazi Flying Saucers & the Dawn of the Flying Saucer

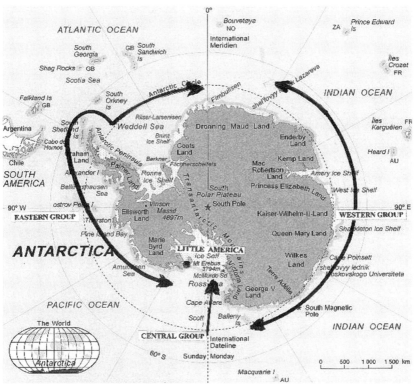

Chart showing the military movements of Operation Highjump.

continent in January 1946, the task force split into four groups that formed what was basically a double pincer movement around the German base in Neuschwabenland, which had been established in 1939 before the war broke out.

Byrd's part of the task force landed south of Neuschwabenland and named the base Little America IV. Once that was established, the four military groups approached Neuschwabenland from four sides. But shockingly, after a little over a month of military maneuvers, the task force was withdrawn and many of the ships, planes and much of the heavy equipment were left behind in what seemed like a mad panic to retreat from the continent. The operation was abandoned a full six months earlier than scheduled.

Almost immediately, rumors began to circulate within the U.S. military that Byrd's task force had met heavy resistance from what were described as "otherworldly" disk-shaped flying vehicles carrying German insignia. According to these accounts, when the task force tried to land in Neuschwabenland, they were

Hidden Agenda

Artist's depiction of a Nazi "Haunebu" flying saucer in Antarctica.

attacked by these flying saucer-like vehicles.

Admiral Byrd himself threw gasoline on these rumors when he gave an interview to International News Service correspondent Lee van Atta aboard the *USS Mount Olympus* after the expedition where he stated that he "didn't want to frighten anyone unduly" but that it was "a bitter reality that in case of a new war the continental United States would be attacked by flying objects which could fly from pole to pole at incredible speeds." Supposedly, these comments come from a story published in the Wednesday, March 5, 1947 edition of a Chilean newspaper called *El Mercurio*:

> Adm. Byrd declared today that it was imperative for the United States to initiate immediate defence measures against hostile regions. The admiral further stated that he didn't want to frighten anyone unduly but that it was a bitter reality that in case of a new war the continental United States would be attacked by flying objects which could fly from pole to pole at incredible speeds. Admiral Byrd repeated the above points of view, resulting from his personal knowledge gathered both at the north and south poles, before a news conference held for International News Service.

Nazi Flying Saucers & the Dawn of the Flying Saucer

It's really quite fascinating to read Admiral Byrd's comments today, because they seem like a clear confirmation of the Highjump rumors. However, no one has been able to confirm the publication of the interview in the years since, even though *El Mercurio* certainly exists and some online claim to have the original clipping, which would have been in Spanish.

But what powered these flying saucer-like vehicles? Did these "Haunebu" aircraft, as they were known, use advanced versions of Die Glocke as their main power source? Or are other rumors, e.g., that Byrd and his task force encountered alien beings living in the underground caverns of the Earth, the true story?

I favor the former. The logical extension of the Nazi Bell technology would be to view it as a reactor of sorts, and place it in the middle of a disk shaped, Repulsine based craft. Just a few such craft could easily defeat a task force armed only with conventional weapons, such as guns and bombs and the like. Also, the field generated by a Nazi Bell type reactor might also act as a kind of shield or force field, making the craft impervious to conventional weapons.

There is no question that Operation Highjump remains shrouded in mystery. But there are other events from the same era that may shed light on U.S. government secrecy at the dawn of the national security state.

When you look at the ME-262 jet fighter, the V2 rocket, the Repulsine and the Nazi Bell, it's obvious that the Germans were decades more advanced than the Allies in aviation technology. This was primarily because the Allies were stuck in the Einsteinian dead end of quantum mechanics, Newton and relativity, while the Germans were more interested in rotational physics and combining the mystical elements of the ancient writings with modern technology. This did not stop the U.S. from attempting to develop the technology that they *had* captured from the Germans, however.

The first major incident of the modern flying saucer era was the Kenneth Arnold sighting near Mt. Rainier, Washington on June 24, 1947, just a few months after Operation Highjump. But while it started the modern UFO era, it may have been an example of advanced German technology rather than aliens.

Hidden Agenda

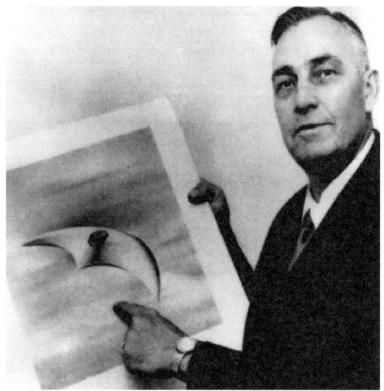

Kenneth Arnold and his crescent shaped "flying saucer."

Kenneth Arnold's sighting took place as he was flying a small aircraft near Mt. Rainier in the state of Washington. What Arnold described was a squadron of six to nine disk shaped or crescent shaped objects "skipping" along at high speeds between him and the mountain. He described their motion as skipping and wobbling, and said it reminded him of teacup saucers skipping along the water. The press seized on this description, came up with the term "flying saucer," and it stuck.

When he had an artist draw up the craft he saw, they bore a striking resemblance to a flying wing style aircraft, which was virtually unheard of in 1947. Arnold's description in fact was almost a dead ringer for a secret Nazi flying wing aircraft that had been captured just after the war in 1945.

Near the end of the war, the Nazis were on the verge of deploying an even more impressive aircraft than the ME-262 jet fighter. It was called the Horton HO-229 and it was the world's first flying wing stealth fighter-bomber. The visual similarity between

Nazi Flying Saucers & the Dawn of the Flying Saucer

Kenneth Arnold's "crescent" (above) and the Nazi HO-229 fighter-bomber (below).

what Arnold described and the HO-229 is pretty remarkable.

The HO-229 flying wing was an experimental prototype flown by the Germans near the end of the war. The one that was captured by the Allies was an incomplete third prototype, and there were plans for three more versions of the aircraft already drawn up. One of the main differences between the V3 prototype that was captured and the previous two was that it was significantly larger, and had provisions for adding a second pilot. This would have had the effect of moving the center of gravity aft, thereby increasing the stability of the airframe.

One of the key clues in Arnold's report was the wobbling

Hidden Agenda

Allied mechanics with the captured HO-229 prototype in 1945.

and skipping motions he noted. According to the Associated Press:

> He said they were bright, saucer-like objects--he called them 'aircraft.' ...He also described the objects as 'saucer-like' and their motion 'like a fish flipping in the sun.' Arnold described the objects as 'flat like a pie pan.'

Also to AP, he was quoted as saying "they appeared to fly almost as if fastened together—if one dipped, the others did, too. "They were shaped like saucers and were so thin I could barely see them..." he told United Press. In an interview on June 27 with the *Portland Oregon Journal*, he said: "They were half-moon shaped, oval in front and convex in the rear. ...There were no bulges or cowlings; they looked like a big flat disk."

Arnold said that the objects weaved "like the tail of a Chinese kite." In his official report to the Army Air Force (AAF) he also added the following description:

> What kept bothering me as I watched them flip and flash in the sun right along their path was the fact that I couldn't make out any tail on them, and I am sure that any pilot would justify more than a second look at such a plane.

He later told Edward R. Murrow of CBS News: "I assumed at the time they were a new formation or a new type of jet, though I was baffled by the fact that they did not have any tails." He also added:

These objects more or less fluttered like they were, oh, I'd say, boats on very rough water or very rough air of some type, and when I described how they flew, I said that they flew like they take a saucer and throw it across the water. Most of the newspapers misunderstood and misquoted that too. They said that I said that they were saucer-like; I said that they flew in a saucer-like fashion.

According to *Wikipedia*, Arnold was later hired by a local magazine to investigate the June 1947 Maury Island UFO incident.

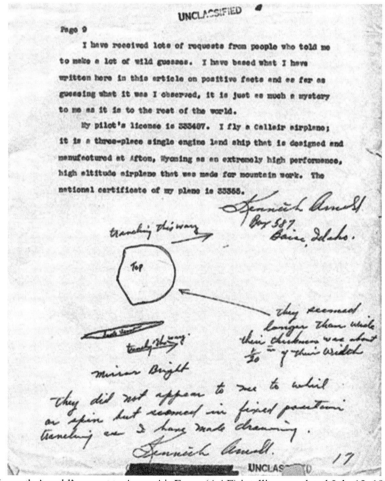

Kenneth Arnold's report to Army Air Force (AAF) intelligence, dated July 12, 1947, which includes annotated sketches of the typical craft in the chain of nine objects.

Hidden Agenda

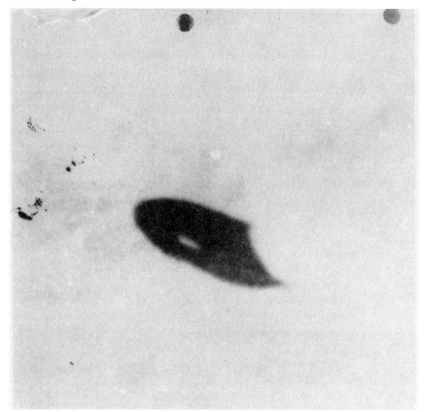

The "Rhodes object" as it appeared in the *Arizona Republic* in July 1947.

Although he eventually turned the investigation over to the Army Air Force, during the course of his discussions with the same two AAF intelligence officers who had interviewed him on July 12, Arnold revealed that one of the nine objects he had seen was different, being "larger and shaped more like a crescent coming to a point in the back." Arnold was then shown the so-called "Rhodes photos" of a crescent-shaped object over Phoenix. Arnold deemed them authentic because of the similarity to the crescent shaped vehicles he had seen.

Clearly, to my aerospace engineer's mind anyway, what Arnold was describing was what we know today as a "flying wing" aircraft. It is possible they were actual flying saucers, but I don't think so for a number of reasons. Arnold's description and the illustration is almost a dead ringer for the HO-229.

One of the advantages of flying wing designs is that they have very little drag and generate tremendous lift, making them

Nazi Flying Saucers & the Dawn of the Flying Saucer

very maneuverable and fast. But the reason they were never really successful was that such an airframe is also inherently unstable and difficult to control. In fact, it wasn't until the development of computer controlled fly-by-wire fiber optic technologies in the 1980s that we were able to get such a design to work, in the form of the B-2 Stealth Bomber.

Now, when you read Arnold's reports, it's clear that the "wobbling and skipping" motion he's observing is right in line with the behavior of an unstable flying wing aircraft. What I think is that the Allies brought back the HO-229 prototype along with the plans, and built a bunch of them to experiment with. If you look at Arnold's drawings, it's clear the cockpit has been moved back relative to the prototype, and this would be done to improve lateral stability.

Also, the area where Arnold had his sighting is right by McChord Air Force base, near Tacoma. Today it's a very well populated area, but at the time, it was almost completely isolated and would have been a great place to assemble and test a small squadron of experimental jet fighters.

Arnold's sighting set off a wave of flying saucer sightings all over the U.S. These sightings included the aforementioned Maury

The Arnold crescent shaped craft.

Hidden Agenda

Island incident, sightings near Cincinnati, on Vashon Island near Seattle and the Rhodes objects near Phoenix. This wave of activity came to a head two weeks later in early July 1947.

About two weeks after Arnold's sighting, around July 4, 1947, something crashed in the desert near Roswell, New Mexico. Interestingly, this craft was also described as disk or crescent shaped, almost exactly like the Arnold report and the HO-229. It's possible that after Arnold's highly publicized sighting, the Army Air Force decided that McChord wasn't quite secure enough, so they transferred the entire squadron of HO-229s to Roswell and the 509th bomb group, the only nuclear armed air group in the world at that time. Roswell Army Airfield had the advantage of being even more remote than McChord. Most likely what happened is that one of the HO-229s went down on Mac Brazel's ranch, and that was the debris he found.

Of course we all know what happened next. The air base issued a press release claiming they had captured a wrecked "flying disk," only to retract the story a few days later under pressure from commanders at Wright-Patterson airfield in Ohio. The next day, the base commander General Ramey staged a press conference with Major Jesse Marcel, who had recovered the debris and asserted that it had just been a misidentified weather balloon. Decades later, Marcel retracted his official testimony and declared that the weather balloon debris was not what he had recovered at

The *Roswell Daily Record* announcing the capture of a flying saucer, July 6, 1947.

the crash site.

Marcel described some very exotic metals with highly unusual characteristics and also some strange symbols on some of the beams and structures. These sound an awful lot like the "runic hieroglyphs" found on the Nazi Bell and later the Kecksberg crash vehicle.

After operation Highjump and the capture of the HO-229, the U.S. *must* have known how far behind the Germans we were technologically. So my guess is they tried to understand the function of these runic symbols on the Nazi Bell and other craft, and probably used them on HO-229 flying wings they were testing. It would also follow that the Germans would have been well ahead of us in material technology and exotic metal alloys

Hermann Oberth (center) and Wernher von Braun (center-right) and their military handlers in the 1950s.

Hidden Agenda

as well. This would account for the strange metals that Marcel reported handling.

While the Nazi flying wing test aircraft scenarios make sense, there continues to be the possibility, expressed by many, that the Germans had outside "help" in developing these new technologies.

Wernher von Braun was once asked if he had anything to do with the Roswell crash, and according to several sources, he said he and some of his German colleagues, who were ardent Nazis brought illegally to the United States under Project Paper Clip, were indeed shown the crash site and debris, along with alien bodies. He was also heard to say that the explanation for the advanced state of German physics and aerospace development was that the Nazis had received help from some outside agency, implying aliens. Hermann Oberth, von Braun's predecessor and mentor, also explicitly expressed the same views:

"We cannot take the credit for our record advancement in certain scientific fields alone; we have been helped." When asked by whom, he replied: "The people of other

Wernher von Braun (3rd from right) and his Nazi cohorts in Ft. Bliss, Texas, shortly after the war.

Nazi Flying Saucers & the Dawn of the Flying Saucer

worlds." In another interview, he was quoted as saying: "It is my thesis that flying saucers are real and that they are space ships from another solar system. There is no doubt in my mind that these objects are interplanetary craft of some sort. I and my colleagues are confident that they do not originate in our solar system."[4]

It's important to remember that von Braun and most of the other scientists brought over in Paperclip were not just loyal Germans, but ardent Nazis. Von Braun was friends with Henrich Himmler and a member of the SS. In fact, there are photos of Von Braun and his cohorts in various locales with swastikas and what look like flying saucers on some of the signs they are posing with. So, was the willingness of von Braun and other German scientists to admit that these sightings were alien, really an attempt to cover up Nazi developed technology?

Theoretically, the U.S. government could have used this idea of an alien invasion of our skies to cover up the real truth, that the war wasn't over and that the Germans were well ahead of us in military technology. But there are other indications that the origin of some of the technology that came out of the postwar era, like transistors and fiber optics, actually came from an extraterrestrial source.

Most people don't know this, but prior to Roswell there was another, far lesser known flying saucer crash in 1942 in Cape

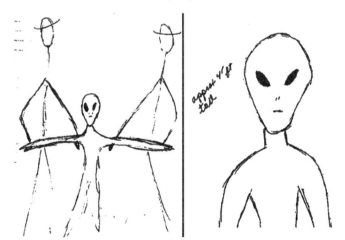

Sketch by Reverend William Huffman of the beings he performed last rites on near Cape Girardeau, MO in 1942.

Hidden Agenda

Girardeau, Missouri. According to the story, a local preacher, Southern Baptist Reverend William Huffman, was called out one night to perform last rites over the victims of what he was told was an airplane crash. When he arrived, he found that the "airplane" was a disk shaped craft, there were three non-human victims of the crash that looked like the typical "Greys," and that the craft had all kinds of these same runic hieroglyphs all over it.

Local fire department records show that there was a fire truck dispatched to the scene on the night of the accident, and that the firefighters were forced to sign military nondisclosure agreements about what they saw there. Later, Dr. Bob Wood, a physicist and documents expert who validated some of the famous MJ-12 documents, discovered documents discussing the Cape Girardeau crash and what was found among the wreckage.

Dr. Wood, who was probably the world's foremost authority on UFOs at the time, having been assigned to investigate the phenomenon for the Douglas Aircraft Company, found documents in the National Archives referring to the Cape Girardeau incident. They described what are clearly known today as fiber optic cables running all through the ship. This discovery was the inspiration for the reverse engineering of the fiber optic technology that would not appear in public until almost 30 years later.

Ironically, it was this same fiber optic technology that made

Los Angeles Times photo of the craft from the "Battle of Los Angeles."

Nazi Flying Saucers & the Dawn of the Flying Saucer

LA Times article on the "Battle of Los Angeles."

fly-by-wire systems possible in advanced aircraft, and ultimately led to successful deployment of the flying wing aircraft more than 40 years after the Horton HO-229 was first tested.

There is one other famous historical incident which fits into this narrative that may now be explained in terms of captured alien or hyper-advanced Nazi technology—the so-called "Battle of Los Angeles" in 1942.

On February 24, 1942, just a few months after Pearl Harbor, Los Angeles air defenses were put on high alert because radar was tracking a number of unidentified craft in the area. Then about 1:00 AM, a large aircraft was spotted flying slowly over Long Beach and anti-aircraft batteries all the way up to Santa Monica fired at it for over three hours as it passed up the coast. Thousands of shells were fired, and somehow, the craft remained impervious to attack, almost taunting the military gunners. The object then turned out to sea and was fired upon by naval forces, to no avail.

The next day, the U.S. military declared that the object was a misidentified weather balloon, and that its soldiers had fired hundreds of shells at it because of "war jitters." None of this sat well with local authorities, who mocked the explanation and alleged a cover-up.

If this was a weather balloon, why wasn't it brought down by the anti-aircraft fire? There is a famous *Los Angeles Times* photo showing what appears to be a solid, disk shaped craft in the spotlights with explosive shells going off all around it. If it was a weather balloon it would certainly have been hit and brought down. But this object seemed unaffected by the attack. Witnesses claimed there were several direct hits. But the object just kept moving. In fact the photo clearly shows a disk shaped object, much like the Nazi Haunebu saucers, caught in the spotlights with shells going off all around it.

One of the theoretical by-products of a torsion anti-gravity field that the Bell could produce would be that it could also act as a kind of protective force field around the craft, rendering it invulnerable to conventional attack by guns and cannons. It certainly looks like this is what we're seeing in these images.

It is also possible that the craft was some other kind of vehicle. Some investigators have pointed out that the distinctive "dome" atop the disk might actually be a flash from an exploding shell, making the craft appear to be less like a Nazi flying disk and more like similar flying disks spotted all over the world.

If you look at the comparison to other flying saucer photos of the period, it does bear a strong resemblance to them. It's possible that the LA saucer wasn't Nazi in origin, but possibly extraterrestrial. On the other hand, I can see the Nazis using this event as an opportunity to test their newly created force field technology. And the event did take place only a few months after Germany's declaration of war against the United States in December 1941.

What is clear is that almost immediately after the Arnold sighting, the Roswell crash and Operation Highjump in 1947, there was a profound shift of power within the United States. One that took us in a decisively dark direction.

(Endnotes)
1 Farrell, Dr. Joseph P. (2004). *Reich of the Black Sun: Nazi Secret Weapons and the Cold War Allied Legend.* Adventures Unlimited Press. ISBN 1-931882-39-8.
2 Farrell, Dr. Joseph P. (2004). *Reich of the Black Sun: Nazi Secret Weapons and the Cold War Allied Legend.* Adventures Unlimited Press.
3 Cook 2001, *The Hunt for Zero-Point*, p. 192
4 http://www.warriorforum.com/off-topic-forum/397611-dr-hermann-oberth-helped-aliens-famous-quotes-about-ufos.html

Chapter 3
MJ-12 and the National Security State

Within a decade of the events of 1947, we see the rise of what is now known as the national security state, sort of a government within the government, in the United States. The CIA was formed just a few months after Roswell, followed quickly by the NSA (which is used to spy on Americans), DARPA (the Defense Advanced Research Projects Agency), the National Reconnaissance Office, and a whole plethora of sub-agencies. According to leaked documents, the Majestic-12 organization was also formed at this time to oversee the flying saucer problem.

Page one of the Eisenhower briefing document.

In 1984, UFO researcher Jaime Shandera received an envelope containing microfilm which, when developed, showed images of eight pages of documents that appeared to be briefing papers describing "Operation Majestic Twelve," or MJ-12. The documents were apparently from a 1952 national security briefing for then president-elect Eisenhower.

The documents indicated that a flying saucer crash had taken place in the New Mexico desert near Roswell, New Mexico, in 1947, just as the legends claimed. They briefed the president on the necessity for creating MJ-12, and listed the members, including the director of the CIA, Admiral James Forestall, and prominent scientists like Dr. Donald Menzel (Carl Sagan's mentor at Harvard) and Dr. Vannevar Bush, President Truman's scientific advisor. The briefing memo describes the aliens, which it refers to as "Extraterrestrial Biological Entities," or "EBEs," as being small, large-headed Grey aliens and states that they are from the binary Zeta Reticuli star system. Their intent is not made clear in the documents but hostility is certainly not excluded as a possibility.

The documents also purported to reveal a secret committee of 12, supposedly authorized by United States President Harry S. Truman in 1947, which handled all matters pertaining to aliens. They explained how the crash of an alien spacecraft at Roswell in 1947 had been concealed, how the recovered alien technology could be exploited, and how the United States should engage with extraterrestrial life in the future. The authenticity of these explosive

The so-called "Twining memo," to Air Force General Nathan Twining requesting his presence at an MJ-12 meeting in 1954.

MJ-12 & the National Security State

documents was quickly challenged, but document experts within the UFO community proved that they were consistent with other documents of the period.

Physicist and documents expert Dr. Bob Wood was assigned to investigate the UFO phenomenon by the Martin aerospace corporation in the 1950s. He concluded at that time that UFOs represented a real, physical mystery. He found the MJ-12 documents to be valid and stated, "There's no question in my mind that they are an authentic briefing memo for president-elect Eisenhower."

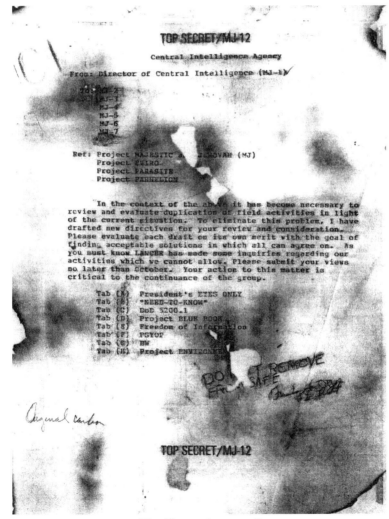

The "burned memo."

Hidden Agenda

Other researchers, like nuclear physicist Stanton Friedman, found other documents which referenced the MJ-12 organization. In a search of the National Archives, he and Shandera later found the so-called "Twining memo" referencing a 1954 meeting of MJ-12. This all but confirmed that organization exists, or at least existed at one time.

Another confirming memo referring to MJ-12, the so-called "burned memo," surfaced even later. It warned that "Lancer," JFK's secret service code name, was inquiring about their activities and "this we cannot allow." It was an ominous warning.

Later, another even more disturbing technical manual emerged, entitled the *"Majestic-12 Special Operations Manual."* It was subtitled "Extraterrestrial Entities and Technology, *Recovery and Disposal.*" Basically this was a "how to" manual for dealing with a crashed alien spacecraft and possible survivors. The purpose was to provide instructions to military recovery units about the background of this top secret program and how to handle the extraterrestrial craft parts, Extraterrestrial Biological Entities (EBEs) and related propulsion and weapons systems while deceiving the public and media saying that nothing important had crashed. Although it came under heavy attack by critics and debunkers, Dr. Wood gave the MJ-12 Special Operations Manual a 97% probability of being authentic. The Special Operations manual is dated 1954 and describes specifically how to handle wreckage and EBEs, be they dead or alive. It talks about how the beings seem to be avoiding contact with our species, and says, as of 1954 anyway, there is no indication that they bear hostile intent towards humanity.

The manual also describes the four basic types of UFOs. The first are the small disk shaped craft, anywhere from 30 to 150 feet in diameter. It describes these as "scout ships," because they contain accommodation only for small crews and have no apparent provisions. At least none were ever found on the disks that crashed.

The second type of alien craft mentioned in the manual is the very large "fuselage" or cigar shaped craft. These are rarely seen at low altitude and sometimes have windows. It's speculated in the manual that these are carriers of some type, possibly the base ships

MJ-12 & the National Security State

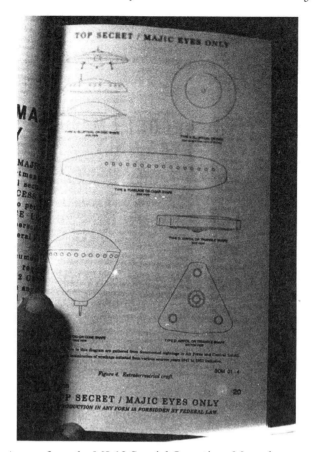

A page from the MJ-12 Special Operations Manual.

from which the smaller scout ships operate.

The third type of alien spacecraft listed is the ovoid, or acorn shaped craft. Their function isn't well understood but they're said to have tripod landing gear and range in size from 20 feet across to up to 100 feet. This may be the type of craft witnessed by police officer Lonnie Zamora in New Mexico in 1966.

The fourth and final type of alien craft mentioned in the Special Operations Manual is the triangle shaped craft. These appeared in the 1950s after the initial wave of flying saucer sightings in the 1940s. Sightings of the triangles are pretty rare, but they have been part of some very prominent cases, like the Phoenix Lights in 1997.

The Phoenix Lights are probably the most famous of the triangle UFO sightings. On March 13, 1997 hundreds of witnesses

Artist's depiction of the Phoenix Lights craft.

all around the city of Phoenix Arizona saw multiple triangle or boomerang shaped craft travel soundlessly over the city. They were so enormous that witnesses said they blotted out huge portions of the stars in the night sky. Later that evening, in an attempt to debunk the sightings, the Air Force sent up parachute flares on the shooting range just over the mountains beyond the city. To this day that's what the Air Force insists people saw. The story didn't take.

I've worked in aerospace for more than two decades, and the general consensus on these kinds of sightings is that if it looks like a disk, it's probably one of theirs. If it looks like a Dorito, it's probably one of ours.

The Special Operations Manual is really the first place that triangular shaped craft are identified as extraterrestrial in origin.

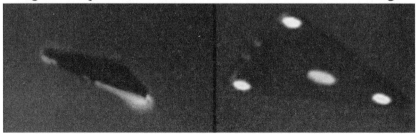

Photos of two triangular shaped UFOs. Are they "theirs," or "ours?"

MJ-12 & the National Security State

The A-12 Avenger.

There could be some confusion here, because the United States is known to have been developing triangle shaped aircraft for quite some time.

The A-12 Avenger was a triangle shaped stealth fighter-bomber that was nicknamed "The Dorito" by the design teams. It was cancelled by Secretary of Defense Dick Cheney in the early 1990s because it was overweight and over budget. There's an old joke in the business that says "if it's overweight and over budget, that's how you know it's an airplane." But it shows the U.S. had an interest in developing air craft with a triangular planform. This only adds to the confusion when trying to differentiate between alien spacecraft and highly advanced human technology that may have been developed in secret.

Of course, the various MJ-12 documents have been attacked and disputed by a number of critics on many grounds. The argument has been made that the presidential briefing document is a fake because President Truman's signature on the executive order creating MJ-12 is identical to his signature on another unrelated document. But back in those days the president and other officials used a device called a pantograph. It's a mechanical device that connects several pens to a pen the president is holding, allowing him to sign multiple documents in one, identical motion, so this argument doesn't really hold up.

And if the MJ-12 documents are a hoax, as many critics like the long dead debunker Philip J. Klass flatly asserted, why hasn't somebody come forward to take credit for the hoax? Isn't 30 years kind of a long time to stay silent? Isn't the point of a hoax to eventually take credit for it so you can show people how clever

Hidden Agenda

you are? If the MJ-12 documents are a hoax, what's the point of the hoax? And isn't it a bit overly elaborate, with all these reams and reams of documents? Why go to all that trouble and then not take credit for it?

As a researcher, I'm less concerned with the authenticity than I am with the content. If the MJ-12 documents are fakes, then they are very professionally created fakes that most likely are government disinformation. That has a whole bunch of implications on its own. But what if they're real? What does that mean about the state of our world today?

It's hard for me to reconcile the conclusions of the presidential briefing documents with the reality of the UFO phenomenon. The documents say the aliens don't seem interested in contacting us, but after the Eisenhower years the incidence of abductions and direct encounters skyrocketed. You have the Betty and Barney Hill case, the Sgt. Schirmer case in Nebraska, the Zamora case and the Pascagoula abductions. These cases and many like them certainly indicate a hostile intent on the part of at least some of the aliens. It's possible that after the MJ-12 presidential briefing was written, things changed between the aliens and humanity, that the relationship took a new, darker turn. It's also possible that the visitors never had a benevolent agenda in the first place.

While debates persist about the reality of MJ-12, there is no question that the UFO problem was addressed at the highest levels of government during the late 40s-early 50s flying saucer flap. On September 23, 1947, just weeks really after the Roswell incident, Army Air Force General Nathan Twining wrote a memo to his boss, Brigadier General George Schulgen about the "Flying Disks." Twining indicated that, "The phenomenon reported is something real and not visionary or fictitious."

What's interesting about Twining is that not only was he listed as an original member of the MJ-12 organization—there's even that memo to him telling him to attend a meeting of MJ-12 at the White House—but it is also true that he was in the Roswell area around the time of the crash incident. He even cancelled a scheduled visit to Boeing a few days later because of "something important" that had come up in New Mexico he had to deal with. He headed the Air Materiel Command at Wright-Patterson Air

MJ-12 & the National Security State

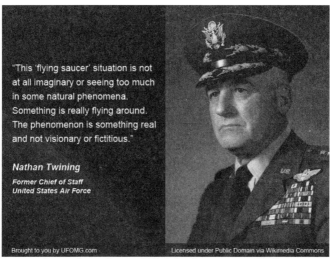

Quote from General Twining's memo on UFOs.

Force Base, where the Roswell wreckage was sent and which was the home of the Air Force Foreign Technology Division. So for him to declare that the phenomenon was "real" was quite a statement.

There had also been a series of alarming UFO encounters in 1947 near Muroc Air Force base in California, known today as Edwards Air Force base, in which Muroc pilots had several near misses with incredibly fast-moving and evasive "silvery disk shaped" aircraft. In response, all Air Force pilots were required to report any such sightings to personnel at Wright-Patterson field, and were told not to discuss the incidents. This led to the first formal U.S. government investigation of flying saucers, code named Project Sign.

After about a year, Project Sign was replaced by Project Grudge, which took the attitude that all UFO sightings should be debunked and dismissed as quickly as possible. After a few years, after declaring that *all* the cases it had investigated were easily explainable, Grudge came to a close. Some of the Air Force high brass were so dissatisfied with its conclusions that they shut down Grudge and replaced it with Project Blue Book in 1952. And right after that is when all hell broke loose...

On July 20, 1952, a major event took place over the United States capital of Washington, D.C. Air traffic controllers at Washington National Airport spotted seven objects about 120

Hidden Agenda

miles out moving rapidly toward the city from the south. The objects were not authorized air traffic and they were not following conventional flight paths.

The air traffic controller who first spotted the blips alerted his boss, Harry Barnes, a senior air traffic controller at the airport, who also tracked the objects. Barnes said that he knew immediately that a very strange situation existed. Their movements were completely radical compared to those of ordinary aircraft.

After confirming that the radar was operating normally, Barnes called the backup radar center and discovered that not only were the objects visible on the radar there, the operator also told him that he could see one of the objects from his control tower. The operator, Howard Cocklin, told Barnes that he could see a bright orange object, moving erratically. Right about that time, the radar screen suddenly erupted with multiple new contacts.

When these radar targets slowed down and began moving over the Capitol Building and the White House, Barnes became alarmed and called Andrews Air Force Base, about 10 miles from National Airport.

The officers at Andrews said they had nothing on their radar, but they did confirm visual sightings of the aircraft. One airman reported an object which appeared to be like an orange ball of fire, trailing a tail, and said it was unlike anything he had ever seen before. As the airman tried to alert the other personnel in the tower, the strange object "took off at an unbelievable speed."[1]

Meanwhile, another person in the National Airport control

UFOs over Washington, D.C., July 1952.

tower reported seeing "an orange disk at about 3,000 feet altitude." A pilot sitting in the cockpit of his DC-4 on the runway, waiting for permission to take off, then reported to the tower that six "silver, tailless, fast-moving lights" were passing over his aircraft. The tower confirmed them on radar and almost the instant this was reported to the pilot, the objects zipped off at high speed and disappeared from the scopes.

By this time, about an hour into the sighting, Andrews' radar was also picking up multiple objects, and a call was made to scramble Air Force fighter jets to investigate.

Right about this same time, Air Force witnesses reported seeing multiple objects that appeared to hover and stand still, then make an abrupt change in direction and altitude. These maneuvers would have been impossible for any conventional aircraft to make.

Some witnesses said that the objects appeared to be balls of orange light; others reported massive silvery disk shaped objects. At one point, all three radars, the two at National Airport and the one at Andrews, triangulated and tracked a solid object hovering over a radio beacon, only to have it abruptly disappear from their screens simultaneously. It's important to note that it did not fly off at high speed. It simply *disappeared*.

By this time, panic had ensued with the objects buzzing all over the capital and violating restricted airspace. Two F-94 fighter jets which had been scrambled earlier from an air base in Connecticut arrived on the scene about 3 AM. When they entered the airspace over Washington, D.C., *all* of the intruder aircraft simply disappeared from the radar scopes. As soon as the jets departed to refuel, the flying objects suddenly reappeared and started buzzing the capital again. This cat and mouse game repeated itself over the next several hours, until dawn came and the aerial objects left for good.

The reaction from the press the next day was explosive. They splashed sensational headlines all over the place. But there was no official reaction until the next weekend, when the objects returned again.

The following weekend of July 26th-27th, the events were repeated with radar returns, visual sightings and even one frightening direct encounter.

Hidden Agenda

This time the fighter jets were waiting, and they were quickly scrambled when a stewardess on a National Airlines flight reported strange lights around her aircraft. When they arrived on the scene, one of the F-94 jets broke off to chase a group of four objects, only to have them suddenly change course and surround his fighter. When the pilot radioed in for instructions from the tower, he was met with stunned silence.

By this time, there were unknown objects being tracked in every sector, with some of them making right angle turns at speeds calculated as being in excess of 7,000 miles per hour! An additional two F-94s were scrambled later and one of them chased a bright light which sped away from him and then simply disappeared.

The next day, the newspapers were once again abuzz with news of the night's events. It was widely reported that the Air Force had issued "shoot down" orders should the craft reappear. Even President Truman made public comments about the events, asking USAF Captain Edward Ruppelt, the head of the Air Forces' newly formed Project Blue Book, to look into the sightings. Before he could issue a report, the Air Force held a sensational press conference headed by the same General Roger Ramey who had quashed the Roswell incident in 1947.

The Air Force said at the press briefing that the objects were "mirages" caused by "temperature inversions" and that the radar operators had simply mistaken them for solid, flying objects. However, the radar personnel said that they were used to weather related returns and "were paying no attention to them." They reiterated their position that all those present in the radar room were convinced that the targets were most likely caused by solid, metallic objects.

With no real explanation for the incursions over Washington, most reporters and the military moved on, preferring to ignore the strange events. But the public remained fascinated.

For years, the conventional wisdom has been that the events in Washington, D.C. on those two summer weekends were visitations by alien spacecraft, performing impossible maneuvers with otherworldly technology. But it has never been satisfactorily explained or understood.

It has been suggested that after the Cape Girardeau and

MJ-12 & the National Security State

Roswell crashes, which were intended to open communications between the aliens and humans, that this was an attempt to make the alien presence known directly to the U.S. government. That it was, for all intents and purposes, the same thing as the proverbial "landing on the White House lawn."

But, given what we now know about the true state of German war technology in the postwar period, perhaps it was something else. Perhaps it was a warning.

The fact is, no one in the U.S. government at the time could have seen these incursions as anything but a serious threat to U.S. national security. It appears that, far from taking it as a benign attempt at interspecies communication from a friendly extraterrestrial race, the US viewed it as something far more worrisome: a shot across the bow from the Nazi remnant in possession of advanced technology operating out of Antarctica.

One thing is certain. After the events of July 1952, which played out almost exactly as Admiral Byrd had warned after Operation Highjump, the United States did not react with calm and consideration.

The truth is, if you look at everything that happened right after the Washington buzzings—the growth of the national security state, the space race, the massive investments in advanced weaponry—it seems like those in the U.S. government made a very big decision at that time. They decided to prepare for war...

***

"We already have the means to travel among the stars, but these technologies are locked up in black projects and it would take an act of God to ever get them out to benefit humanity.... Anything you can imagine, we already know how to do."
– Ben Rich, Director, Lockheed Skunk Works

After the unexplained flyovers of Washington, D.C. in July 1952, the government of the United States seemed to go into panic mode. Within months, new and powerful agencies were created or given bigger budgets, seemingly in response to the events of that summer. After the incursion into the airspace over Washington,

Hidden Agenda

D.C., the U.S. government put on a public face of calm, but behind the scenes they were in a frenzy. Operation Paperclip, the importation of German scientists and engineers, was dramatically accelerated over the objections of President Truman. It's obvious when you look back on it that somebody in Washington decided the U.S. needed to get caught up in the technology race.

Even though the executive order authorizing Paperclip specified that no "ardent Nazis" would be allowed to immigrate under the policy, that was all put aside around 1952. Nazi party members like Wernher von Braun, who was personal friends with Heinrich Himmler, along with Kurt Debus and Hubertus Strughold, who were war criminals, were suddenly deemed acceptable. Some of them were even granted visas through embassies in South America in order to sneak them in through the Mexican border. Most of them had elaborate Nazi pasts that were swept under the rug.

In retrospect, it's easy to see why the U.S. was suddenly willing to bend the rules. They were surrounded on three sides by the enemy. Soviet Russia and her nuclear bombs, a Nazi remnant in Antarctica with their Haunebu saucers, and possibly aliens from above coming to Earth and engaging in who knows what kinds of activities. The result was a vast acceleration onto a wartime footing by the Unites States. Besides Paperclip, the National Security Agency, or NSA, was formed to spy on domestic activities, and the Defense Advanced Research Projects Agency, or DARPA, was created to develop new weapons systems, including what would eventually become the so-called "Star Wars" program. Other agencies and laboratories, like the Defense Intelligence Agency, the National Reconnaissance Office and the Lockheed Skunk Works, which developed the U-2 and SR-71 spy planes and the F-117A stealth fighter, were also quickly brought online.

The U.S. must have had a major inferiority complex at this point, especially given the likelihood that our proudest wartime technological achievement, the Manhattan Project, was probably not ours at all, but the result of captured German technology!

As late as May 1945, J. Robert Oppenheimer, the wartime head of the Los Alamos National Laboratory, wrote to U.S. president FDR and told him that at their present rate of uranium

MJ-12 & the National Security State

enrichment, the earliest the U.S. could expect to detonate an atomic bomb was November of 1945.[2] Yet somehow, the U.S. conducted the Trinity test detonation in Nevada in July and then dropped two atomic bombs, code named Little Boy and Fat Man, on Japan in August 1945. But if the US didn't possess enough uranium to build a single bomb in May 1945, how did they manage to detonate three just a few months later?

It turns out that the U.S. Navy had captured a German submarine, the U-234, with three Japanese officers aboard as well as six barrels of enriched uranium and three atomic triggers, which the Germans were well ahead of us in developing. The original manifest of the U-234 shows that all of this was on its way to Japan, where the Germans hoped they could be assembled into atom bombs to be used to save the war effort. As it turned out, this cargo was instead captured by the Allies and used to construct the bombs that were dropped on Japan. The point is, even our proudest technological achievement of the war was a fraud.

All of these new "alphabet soup" agencies and laboratories appeared to end up under the control of a single, ultimate power: the group known as MJ, or "Majestic-12."

The documents show that MJ-12 was formed in 1947, just after the Roswell incident, as the major, uber-top-secret overseers of the alien problem. It's reasonable to assume that they also held sway over all the other U.S. intelligence agencies and labs, since ultimately they would all fall under the national security umbrella that MJ-12 was created to manage. According to the documents, members included men like Dr. Vannevar Bush, who was the science advisor to presidents Roosevelt and Truman, General Nathan Twining, who controlled the Foreign Technology Division at Wright-Patterson Air Force Base, and Rear Admiral Roscoe Hillenkoetter, who was the head of the CIA.

There are many supporting documents that prove MJ-12 exists, including the verified memos from the National Archives calling members to meetings at the White House, for instance. But the real question is whether MJ-12 was created to deal with the alien problem, the Nazi problem, the Soviet problem, or all three? Ultimately, the reasons don't really matter. What happened next is the real story.

Hidden Agenda

Patents were issued in the early 1970s for something called a "nuclear subterrene," a heated metal drill that would bore its way through solid rock. The idea was that this drill would heat up and liquefy solid rock, and the molten rock would cool and form a glass-lined tube behind the drill as it moved forward. The whole thing was powered by the heat from a compact nuclear reactor, much like the ones onboard modern day nuclear submarines and aircraft carriers.

The only reason you apply for a patent is if you're ready to introduce the technology publicly and commercially, so the subterrene patents imply that these were developed for defense purposes probably 20 years earlier. According to many rumors, using devices like nuclear subterrenes, the U.S. government began to create a vast underground network of tunnels. These tunnels are said to crisscross the western United States, connecting bases like Area 51 in Nevada to laboratories like Los Alamos in New Mexico, and to secret bases in Dulce, New Mexico, S-4 in Nevada, Sedona, Arizona and as far east as Wright-Patterson Air Force Base in Ohio.

So what happened is that MJ-12 became in effect a secret government designed to respond to the Nazi/alien threat in whatever way they saw fit. They had their own budgets, their own battle strategies and most importantly, their own technology programs with the aim of catching up to the Germans and the ETs who may have been coming to Earth. Using these resources, it seems that they split into two factions, one public, and one private. They were in competition with one another to see who could develop the best technologies the fastest. While the secret government side seemed well aware of the public space program, the public side was kept in the dark about the existence of the secret space program.

Up to this point, the presidents, Truman and Eisenhower, had been kept in the loop about the secret government and MJ-12. But it seems that by the time Kennedy took office, presidents were no longer considered "need to know" personnel.

Once these various agencies established a foothold in the mid-1950s there starts to be this very sudden new interest in what are called "field propulsion systems," especially in the work of an American experimental scientist named T. Townsend Brown.

T. Townsend Brown was an American inventor and engineer from a well-to-do family who experimented with high voltage electricity in his basement laboratory. His preeminent discovery was that if you ran high-voltage electricity through a capacitor, the charged object would *move*. Once he realized this, Brown began a series of experiments constructing devices of various sizes and shapes. What he found was that the "flying saucer," or disk shaped test article was the best for reproducing the effect.

In 1929, Brown published an article in *Science and Invention* magazine titled "How I Control Gravitation," in which he explained the effect.

What Brown found was that if you put these specialized capacitors, which he called "Gravitators," on this disk shaped apparatus and pumped about 200,000 volts through it, it *moved* in the direction of the positively charged plate.

He discovered a number of things in his early experiments. First, the higher the charge, the faster and more powerful the effect was. Second, the higher the voltage, the more powerful the effect was, and third, the more massive the test rig was, the greater the movement. He also discovered that the effect was influenced by the positions or alignments of the Sun, Moon and planets. Careful observers will note this is right out of the hyperdimensional physics playbook I outlined in my book, *The Choice*.

After testing in a vacuum chamber proved that the effect was not simply the result of pushing ionized air around the disk,

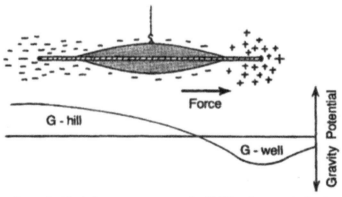

High-voltage electrical charge creates a gravity "hill" at the negatively charged end, and a gravity "well" at the positively charged end. This effectively "warps" space-time and allows a disk shaped craft to "surf" in the positive direction.

Hidden Agenda

Brown realized that there was some kind of connection between electromagnetism and gravity. He called his new science "electro-gravitics." Today, we would simply refer to it as an "anti-gravity" effect.

Even though his results were published and repeatable, Brown met a lot of resistance to his ideas. By the 1920s, Einstein and relativity had taken hold of physics, and that camp said that there was no connection between gravity and electricity, so Brown's device was "impossible" despite the experimental results.

What seems to be happening is that the electrical energy creates a gravitational "well" in the positive direction of charge, and a gravitational "hill" at the negatively charged end. The result is that the disk-shaped craft "surfs" on the crest of a gravity wave that moves constantly in the positive direction. It effectively falls forward on a continuously regenerated gravity wave. What Brown discovered is nothing less than the "warp drive" from Star Trek.

Now, according to the laws of physics, like Newton's third law of motion and the first law of thermodynamics, what Brown's devices did was impossible. They violate the axiom that "for every action there is an equal and opposite reaction" because they produce forward motion without any kind of physical force, like a rocket thruster. They also violate the Law of Conservation of Energy, which states that energy cannot be created or destroyed, but can only change form. What Brown's experiments showed was that these "laws" of physics are simply wrong.

Brown's experiments also showed that a key aspect of Einstein's theory of relativity was incorrect. According to Einstein, the only thing that can distort space-time is mass, because it has an as yet not fully understood connection to gravity. What Brown showed is that gravity and space-time can also be affected by electromagnetism, which relativity says is impossible.

All of the standard physics models are based on a single assumption: that the Universe is a "closed system," and you can't get energy from outside that system. Brown proved that wrong.

Brown spent time at Caltech and other institutions in the 1930s attempting to generate interest in his experiments, but, given the hold relativity and quantum physics had on the scientific world, no one was willing to fund high level research into his theories. Brown continued his experiments on a shoestring and made even more astonishing discoveries.

As Brown continued his experiments privately, he found that certain exotic materials enhanced the anti-gravity effect dramatically. He found that a gravitator made with a barium-titanate dielectric insulating layer greatly reduced the amount of voltage necessary to create forward thrust. In fact, if you were to scale up Brown's designs, travel time to Mars and even the nearest star systems would be greatly reduced.

If you look at power-to-thrust ratios in terms of Newtons per kilowatt, a modern-day fighter jet engine is about 30% efficient. It produces about 15 Newtons per kilowatt of power. By contrast, Brown's first generation electrokinetic thruster produces about *2,200 Newtons per kilowatt,* making it 4,400% efficient. This makes it effectively an "over-unity" device. You're getting far more power out of it than you put in. His second generation design, with the barium-titanate insulating layer, is over 140,000% efficient, generating 70 *thousand* Newtons per kilowatt.

With that kind of power curve, you could send a 100-ton spacecraft—that's 30% bigger than the space shuttle—to Mars in just five days. Using conventional rocket or ion propulsion, the journey would take nine months.

Not only could you make the journey in five days, but the amount of energy required to power the engines is only 3,000 watts, which could be generated by solar cells. At a 70,000 Newtons per kilowatt output the total cost of the 55 million kilometer journey would be about 360 kilowatt hours. At 7 cents per kilowatt hour, the whole trip would cost around $25.

By contrast the fuel cost of a single space shuttle launch is $425 million, or about $3 million per kilowatt hour. If you extrapolate this technology, making the engines more efficient and using compact nuclear reactors as a power source, faster-than-light speeds could theoretically be attained. In fact, some calculations indicate that speeds of up to 200 times the speed of light could be attained, which in Star Trek terms is about warp factor six. At those speeds, it would be possible to travel to Alpha Centauri in about a week.

After World War II, Brown refined his inventions and made

Hidden Agenda

a new proposal to the U.S. military. This time, he was apparently dealing with a more receptive audience.

In 1952, Brown tried one last time to get the attention of military leaders. At a meeting in Washington, D.C. he presented his idea for Project Winterhaven,[3] which was designed to create propulsion and communications systems based on his earlier work. The proposal was very extensive and seemed to spark the first trickle in a wave of interest from the military in anti-gravity technology.

The period from 1952 to 1957 is kind of the "golden age" of anti-gravity in the U.S. Not only was Brown pitching his ideas to industry and military leaders, but pretty much every company in the defense sphere of operations was trying to get in on the act. There are extensive newspaper articles from the period which identify the U.S. government's interest in anti-gravity propulsion, and it's pretty obvious they were trying to play catch up. There are even notices from the Glenn L. Martin aerospace company looking for engineers interested in developing anti-gravity propulsion systems.

While the Navy and other U.S. military agencies rejected Brown's Project Winterhaven proposals publicly, there are indications that internally, they took his work seriously.

During this same period, the U.S. military opened a number of program offices dedicated to the study of anti-gravity effects. One of the most prominent was the Aeronautical Research Laboratory (ARL) at Wright-Patterson Air Force Base. There was also the Research Institute for Advanced Study, or RIAS, which shared information with the ARL. All of this research lasted for about five years in the middle of the decade, then it suddenly all went silent.

It doesn't take the proverbial rocket scientist to figure out what happened.

The military invited and encouraged all of these anti-gravity ideas, and then after they got all the detailed proposals, they threw a wall of secrecy up around the whole subject and took it private. They've probably been working on developing the technology in secret for decades now.

One possible application of this technology has been identified by Dr. Paul LaViolette in the design of the B-2 stealth bomber.

In some of the declassified technical documents relating to the B-2 stealth bomber, reference is made to the fact that the leading edge is electrified. In Brown's own 1965 patents, he describes a method by which jet engines could be supplemented and even replaced by using his methods. The idea was that you would charge the leading edge with positive ions, and then essentially dump the negative ions into the exhaust. By using this method, the B-2 could literally surf a gravity wave, requiring far less fuel and gaining far better performance.

There also seems to be a clear connection between Brown's work and a recent wave of tests of what are being called "E M Drives" that also seem to generate thrust in violation of the so-called "laws of physics."

Recently, tests have been conducted on a number of different designs of what are being called E M Drives. Experimenters have created closed metal boxes called resonant cavity thrusters, into which they bounce high frequency microwaves. Now there's no way in conventional, relativistic terms that such a device can possibly create directional thrust. But when they measure the results, that is exactly what they get.

After initial testing by several private firms, both the Chinese Northwestern Polytechnic University and NASA conducted tests on different designs based on the same concept. The results were astounding.

Both the Chinese and NASA tests came back showing positive, directional thrust, which should be impossible. There's no way, if Einstein is right, that you can bounce microwaves inside a box and develop forward thrust.

These types of "reactionless motors" can be scaled up

A "resonant cavity thruster," or E M Drive.

pretty easily to the kinds of speeds that make interplanetary travel feasible. And since, like Brown's design, they don't use any fuel, the costs associated with say, a trip to Mars are negligible. It's also really interesting that some of the official NASA artwork depicting an E M Drive powered craft shows it riding a gravity wave, exactly like Brown's Project Winterhaven proposal.

There's no question that when you look at the E M Drive you're seeing a further refinement of Brown's electrogravitics work. When you bounce these resonant microwaves inside the conical enclosure, you get a net thrust toward the wider end. This is probably because you're creating an asymmetric charge inside the chamber, and the entire assembly is moving toward the positively charged end.

Another interesting aspect of the E M Drive is that by using superconducting materials on the cavity, thrust could theoretically be increased by a factor of about one thousand, enough to possibly launch a satellite into space. That's pretty good for a device that mainstream scientists say is impossible. There seems to be a separation around this period in which you had these super-secret research labs gathering information and presumably conducting experiments on anti-gravity technologies, and the public space program which was still chugging along trying to develop chemical rockets.

By 1959, most of the field propulsion research had disappeared from the public eye and either been transferred to the black ops world or possibly scrapped. But meanwhile, another anti-gravity field effect was about to be discovered which would change the course of the history of space exploration.

(Endnotes)

1 https://en.wikipedia.org/wiki/1952_Washington,_D.C._UFO_incident

2 Farrell, Dr. Joseph P. (2004). *Reich of the Black Sun: Nazi Secret Weapons and the Cold War Allied Legend.* Adventures Unlimited Press. ISBN 1-931882-39-8.

3 http://starburstfound.org/aerospace/projectwinterhaven.pdf

Chapter 4
A New Front in a Secret War

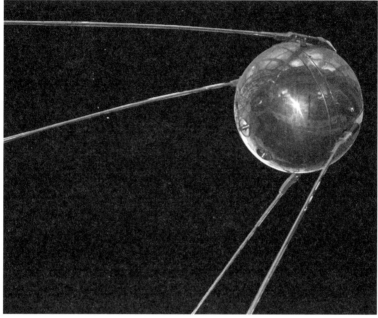

Spuknik 1.

The world was shocked in 1957 when the Russian satellite *Sputnik 1* was successfully launched into Earth orbit. The event sent shockwaves through the U.S. intelligence community, and rocket projects which had been held back or underfunded suddenly got high priority.

The U.S. government basically had two competing programs to choose from at this point. The Navy, and its Vanguard rocket program, or Wernher von Braun and his team of Nazi rocket scientists out of Redstone Arsenal in Huntsville, Alabama. The government preferred the Navy rocket at that point because of the stigma of using German scientists, but when the *Vanguard* rocket blew up on the launch pad in December 1957, they had no choice but to turn to von Braun and his team.

The satellite portion of the Vanguard rocket was actually

Hidden Agenda

tossed into the bushes near the launch pad and began transmitting. The newspapers nicknamed it "Kaputnik" and began mocking the Navy's program. The U.S. had been embarrassed on the international stage and asked von Braun and his team if they could get a satellite into orbit to match the Soviets. Von Braun reminded them he had told them his team could have done it months before *Sputnik*.

On January 31, 1958, the American satellite *Explorer I* was successfully launch atop a modified four-stage Jupiter-C Redstone rocket, rebranded the Juno I for civilian purposes. It successfully achieved Earth orbit, but not without a mysterious "missing time" event that is still unexplained to this day.

At the time, there were only three stations in the worldwide satellite tracking network. Everybody knew exactly how long the thrusters would burn at each stage, and they knew within a very tight window when they could expect the tracking station outside San Diego to pick up the carrier wave signal. It should have come at exactly 12:31 AM on February 1, 1958. It didn't.

Based on a straightforward calculation called the "rocket equation," *Explorer 1* should have been in range by 12:31 AM eastern time. As the minutes ticked by, it became more and more obvious that the spacecraft was lost, and had not achieved orbit. Then suddenly, at 12:42 AM, more than *eleven minutes* overdue, the signal was heard at the Earthquake Valley tracking station near San Diego.

I can't emphasize how impossible this is. Everything about that rocket was well accounted for and understood. Thrust, speed, wind resistance, specific impulse—all that goes into the rocket equation. For *Explorer 1* to have been 11 minutes late is like a car at the Indianapolis 500 arriving two seconds late even though you know the exact speed of the car. It defies the laws of physics.

After the shocking reappearance of a satellite everyone had thought was lost, von Braun and his team quickly scrambled for an explanation. Von Braun suggested in a newspaper interview that they had made "just a slight error in our quick estimate of the satellite's initial speed and period of revolution." Later, NASA historians tried to suggest that the jet stream had boosted the spacecraft into a higher orbit.

A New Front in a Secret War

The jet stream explanation is particularly laughable because the aeronautical scientists knew all about it by 1958, having sent a lot of weather balloons up specifically to study it. Furthermore, the jet stream is a horizontal, easterly flowing wind force, not a vertical one, and if anything would have slowed the rocket down and made its orbit lower, not higher.

The jet stream explanation becomes even more unlikely when you consider just how much of an increase in altitude that extra 11 minutes translates to. The orbit of *Explorer 1* was always expected to be very elliptical, or egg shaped. The calculations had estimated an orbit with a minimum altitude of 220 miles, about the orbit of the Space Shuttle, and a maximum altitude of 1,000 miles. The actual orbit came in at 225 miles, a difference of two and a quarter percent, and 1,594 miles, an increase of almost *600 miles*! That's a 60% higher orbit than they expected, and it's also why the spacecraft was 11 minutes late to the tracking station. I don't know any engineer, including von Braun I'm sure, who actually thinks a 60 percent deviation is only a "slight error" in his calculations!

Von Braun apparently agreed, because almost immediately he began questioning scientists around the world, expressing an

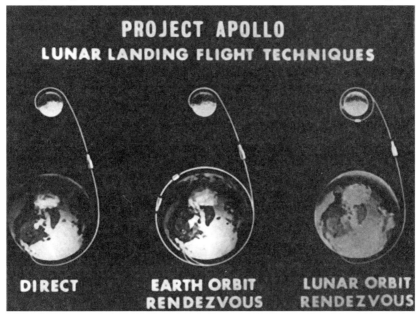

The three proposed methods for traveling to the Moon: Direct Ascent, Earth Orbit Rendezvous and Lunar Orbit Rendezvous.

Hidden Agenda

intense interest in what was then called "field propulsion," or anti-gravity research.

One of his first entreaties was to American physicist Burkhard Heim, who had just given a series of lectures on anti-gravity in Germany and had worked for the Glenn L. Martin aerospace company on its anti-gravity research. He told Heim that without such a "field propulsion system," he couldn't be responsible for the enormous costs of the upcoming Moon programs.

What he meant by "enormous cost" was rocket fuel costs in terms of both money and weight. The accepted method at that time for a manned Moon landing was a technique called "direct ascent," which you see in a lot of 1950s science fiction movies. This is where you launch a whole rocket to the Moon, and then it lands in one piece, takes off again and then lands on the Earth in the same way.

Von Braun found however that by 1958, anti-gravity research had all but disappeared from the scientific literature in the United States and it was nearly impossible to get government agencies or laboratories to talk about it. But when he came across the work of a French Nobel Prize-winning economist named Jacques Allais, he renewed his interest.

Jacques Allais is a French Nobel Prize-winning economist who also liked to dabble in physics (see *The Choice*). During a pair of eclipses over Europe in the 1950s, he conducted experiments with free swinging pendulums. What he found was that before the eclipse, the pendulums swung normally, with the rotation of the Earth. But during the eclipses they suddenly started swinging very rapidly *backwards*, *against* the rotation of the Earth. When he repeated the experiments a second time, he got the same results. Somehow, the eclipses were affecting the swing of the pendulums in unexpected ways.

This behavior became known as the "Allais Effect" and has been observed numerous times over the years. It violated Einstein's special relativity by showing that instantaneous action at a distance was not only possible, but verifiable. After von Braun talked to Allais, he made arrangements to have Allais' papers published in English-language science journals and brought him to America to meet with NASA scientists.

A New Front in a Secret War

But by then, von Braun had bigger problems. The more satellites that were launched, the more obvious it was that the American space program had a huge problem with navigation and guidance of its rockets. When *Explorer 3* was launched in March,1958, it experienced the same problem, only worse. It attained a maximum altitude of 1,750 miles, more than *750 miles* higher than estimated. The Navy didn't quit on its Vanguard program either, and they experienced the same problem.

When *Explorer 4* was launched in July 1958, it also attained an orbit nearly 400 miles higher than calculated. Something was going on, but what? Any notion that the effect was due to winds or miscalculation was swept aside when the Soviets attempted the next milestone in the heated up space race: targeting the Moon.

In early January of 1959, the Soviet Union launched *Luna 1*, a probe that was to be put on an intercept trajectory to the Moon. Since landing a spacecraft on the Moon was impossible at that time, the idea was to just target and crash the satellite into the Moon, which was still a huge milestone achievement in the space race.

Targeting the Moon should have been relatively easy, even with 1950s technology. But somehow, *Luna 1* managed to miss the target. Basically, shooting the Moon in a vacuum should have been like shooting fish in a barrel. All you have to do is boost the probe into orbit, and then fire the thruster on a trajectory to the spot where you know the Moon is going to be in two days. Since the target is 2,160 miles across, and you know where it is relative to the Earth, and you know to an exact degree what the gravity is of both bodies, it should be about as difficult as hitting a backstop with a baseball—from home plate. But when *Luna 1* arrived in the vicinity of the Moon, it missed by *3,700 miles*, about one-and-a-half times the diameter of the Moon itself! After missing the Moon entirely, *Luna 1* ended up in orbit around the Sun, where it still is today.

Less than two months later, DARPA attempted the same mission with a new satellite designed by von Braun named *Pioneer 4*. Its mission, just like *Luna 1*'s, was to be sent on a collision course to the Moon and transmit data right up until the time of impact.

Hidden Agenda

Pioneer 4 was unique in that it was the first spin-stabilized satellite ever launched. The idea was to use a gyroscope effect as a guidance system making it easier to actually find and hit a target in space, like the Moon. Unfortunately, *Pioneer 4* was an even bigger failure than *Luna 1*. It ended up missing the Moon by over *37,000 miles*, more than 17 times its diameter!

In a JPL report issued after the spacecraft failed to impact the Moon, NASA suggested that the spacecraft had somehow picked up "excess velocity" on its way to Moon. The JPL report said that the spacecraft had achieved a velocity slightly *less* than 1% of the estimate NASA had expected. But somehow, along the way, it acquired additional forward thrust, which caused it to miss the Moon. According to Newtonian laws of motion, that's a physical impossibility!

Things didn't get any better after that. The Ranger program was intended to impact the Moon with a much more complex satellite design that even included a mid-course correction rocket booster. Despite this improved design, *Ranger 3* also missed the Moon by nearly 23,000 miles, or nearly *12 times* the Moon's diameter.

Von Braun must have been pulling his hair out at this point, trying to figure out what was wrong. But a crucial clue was forthcoming, in the form of *Ranger 4*.

Ranger 4 was launched in April 1962 and performed flawlessly until its solar panels failed to deploy. Without solar power, the batteries were quickly drained and the spacecraft became a dead clump of metal on a ballistic trajectory toward the Moon. By all logic, without a working guidance system or the ability to make a mid-course correction, it should have missed the Moon as badly as all the other missions had. But it didn't. It actually impacted the Moon, pretty much exactly where it was supposed to!

What von Braun *must* have figured out at this point was that something in the dead spacecraft was different from what was in all the "live" ones that had problems. It probably didn't take him long to figure it out.

The difference was in the rotation. All of the spacecraft which had shown this anomalous overperformance had major components or subsystems which rotated at high RPMs. In fact, there was a direct correlation between the amount and duration of the spin and the spacecraft's performance.

The *Explorer 1* Juno rocket for instance had a rotating third stage. This was because it had a cluster assembly of solid rocket boosters which had a tendency to fire unevenly. Because of this they rotated the entire third stage at 750 RPM to balance out the thrust. The Pioneer and Ranger spacecraft were both spin stabilized and they also had rotating gyroscopic guidance systems on board. When *Ranger 4* went dead, the gyros stopped spinning and suddenly all their calculations worked, and it actually hit the Moon!

All of this most likely led von Braun to create an empirical solution to the spin-guidance problem. Von Braun must have figured it out at that point because the next two Rangers, *Ranger 6* and *7*, impacted the Moon exactly where they were planned to.

You can also tell that he had found his solution by the fact that he suddenly changed his mind about the direct ascent concept. At a design review just after the *Ranger 6* and *7* missions, von Braun surprised everybody when he came out in favor of lunar orbit rendezvous as the best way to get men to the Moon and back. Up until that time, he had been a staunch advocate of direct ascent as the only way to get to the Moon and back. The reason is obvious. If we couldn't even figure out how to guide a spacecraft on a ballistic trajectory and hit a 2,160-mile-wide target, how could we possibly send men to the lunar surface and back again?

The idea that we could guide and track two spacecraft to land on the lunar surface and then dock in lunar orbit was unthinkable given our track record to that time. The fact that von Braun now felt confident enough to embrace a trip to the Moon involving the docking of two spacecraft in lunar orbit was a crucial development given the changing political climate of the time.

By 1961, John F. Kennedy had taken office and replaced the original NASA director with 33rd degree Freemason James E. Webb. At that point, things seemed to shift regarding MJ-12 and the whole UFO problem.

Unlike Truman and Eisenhower, there's no indication that JFK was informed of or involved with MJ-12. He appears to be

Hidden Agenda

the first president that was frozen out of the organization and its activities. In fact, within a few months of taking office, Kennedy directed very stern words toward secret societies and clandestine intelligence operations in a speech entitled "The President and the Press."

> The very word 'secrecy' is repugnant in a free and open society;" he said. "And we are as a people inherently and historically opposed to secret societies, to secret oaths and secret proceedings."
> We decided long ago that the dangers of excessive and unwarranted concealment of pertinent facts far outweighed the dangers which are cited to justify it. That I do not intend to permit to the extent that it is in my control. And no official of my Administration, whether his rank is high or low, civilian or military, should interpret my words here tonight as an excuse to censor the news, to stifle dissent, to cover up our mistakes or to withhold from the press and the public the facts they deserve to know.

What Kennedy laid out that night was nothing less than a declaration of war on the secret societies, like the Freemasons and the SS, which had infiltrated NASA, the Federal Reserve and the military/industrial complex headed by groups like MJ-12. It seems pretty clear from this speech and other sources that Kennedy was frustrated by his inability to get information on some of the most secret operations that America was undertaking. After the Bay of Pigs disaster, his trust of the CIA and intelligence agencies in general was at an all-time low, and there are indications he began to pressure MJ-12 for information on UFOs and the alien problem. He was met with stone-solid resistance.

There is a leaked document popularly called the "Burned Memo" which I mentioned earlier, which was supposedly pulled from a fire by the grandson of a former intelligence officer.

It's an undated MJ-12 memo which urges the members to come up with a plan of action for some unnamed issue or problem facing the group. In it, there's a very cryptic, chilling line that says, "As you must know Lancer has made some inquiries regarding

A New Front in a Secret War

our activities which we cannot allow." Lancer, of course, was the Secret Service code name for President Kennedy.

The memo makes it clear that Kennedy is poking around the activities of MJ-12, and makes it equally clear that they have no intention of letting him find out what they're up to.

All of this activity around the beginning of the Kennedy administration implies that a major policy shift had taken place behind the scenes in the U.S. intelligence community. Speculation centers on a possible new priority for the secret space program.

My guess is that having taken the anti-gravity research into the deep black world and lost control of it, the public side of the space program (Kennedy, von Braun and guys like that) find that they are hopelessly behind not only the alien technology, but probably the German and even the Russians as well. Having solved the problem of how to navigate to the Moon and back, a major decision is made: rather than develop their own anti-gravity propulsion systems, the quicker solution is to simply go to the Moon, where they will likely find abandoned "Anunnaki" technology, and reverse engineer it, much like was done at Roswell. This then becomes the top priority of the public space program, even if it has a secret, hidden agenda buried within it.

By going to the Moon in this way and bringing back the "technology of the gods," as it were, they could theoretically have cut years off the development of flying saucer technologies.

The so-called Brookings Report, a document commissioned by NASA in 1958 at the dawn of the space age, makes it clear that it is "likely" NASA will find "artifacts" of some kind somewhere in the solar system, and it specifically mentions the Moon and Mars. Given that both the Kennedy White House and the Congress had copies of this report, it seems logical that they viewed bringing artifacts back from the Moon as a way to bridge the technological gap between the public space program and the runaway secret one. By going around MJ-12 and the intelligence communities with the public Apollo program, Kennedy and the civilian elected officials could have effectively dealt themselves into the game with secretive agencies like MJ-12. But this strategy was not without risks—or consequences.

Besides the speeches he gave, Kennedy also did several

other provocative things. In April of 1963, he issued an executive order allowing the United States Treasury, rather than the Federal Reserve Bank, to issue currency backed by silver in the Treasury vaults. These "United States Notes" would have completely undermined the stranglehold the privately owned Federal Reserve held over the U.S. economy and political power.

But that was nothing compared to what Kennedy did later in 1963. In a September speech before the United Nations, Kennedy outlined a proposal for the United States and the Soviet Union to go to the Moon together, in a joint program. This would have required the sharing of not only our most highly secret rocketry technologies, but also we would be sharing whatever we found on the Moon with the Russians. Knowing that we were going to find artifacts, remnants of alien technology, this must have shocked and alarmed everyone inside groups like MJ-12. So by the Fall of 1963 it wasn't a matter of who wanted to kill JFK, but more who didn't want to kill him.

Historical records now show that Kennedy had actually made the offer to go to the Moon jointly with the Soviet Union twice before, and had been rejected both times. But this third time, according to a highly placed source, Khrushchev decided to agree to the idea. Nikita Khrushchev's son Sergei said in a PBS interview that he remembers walking with his father in late October or early November 1963. His father told him that Russia had fallen behind the United States and that they would lose the race to the Moon, so he figured that this time, he had better accept Kennedy's offer. There is no official record that this was communicated to the United States, but we can tell from what Kennedy did next that it must have been.

On November 12, 1963, President Kennedy issued his final National Security Action Memo, number 271. In it, he specifically instructs James Webb, the director of NASA, to open an office to streamline "substantive cooperation" with the Soviet Union in support of his new space cooperation initiative.

Now, there is absolutely no reason Kennedy would do this unless he already had an agreement in place with Khrushchev to go to the Moon together. We're talking about opening an office, gathering information and actually establishing lines of

Right: Wernher von Braun's early identity card at Redstone Arsenal. Von Braun: 00005. Below: A Saturn V rocket taking off with the Apollo 11 crew on July 16, 1969.

The Horten Flying Wing.

Left and Above: Two Soviet illustrations from 1961 of the planned base on the Moon. Below: An early NASA concept for a manned base on the Moon.

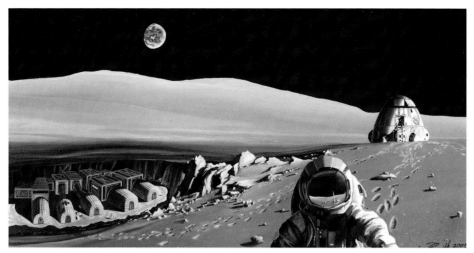

Three concept drawings of a Moon base from the European Space Agency (ESA).

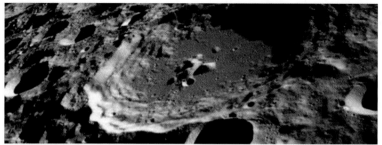
A photo of the crater Gassendi on the Moon with its mysterious domes.

The European Space Agency has plans for similar domes on the Moon.

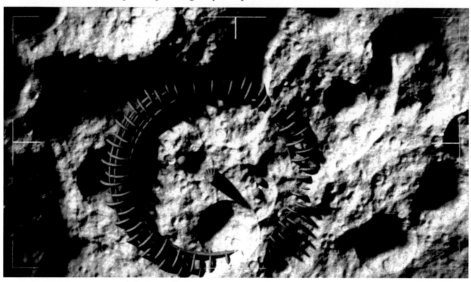
A controversial photo from a YouTube video on secret bases on the Moon.

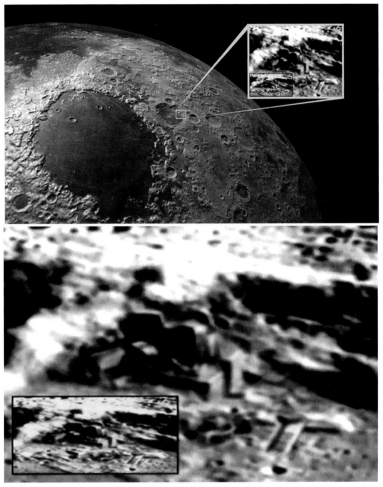

Above: Chinese media in February 2012 suggested that a Chinese Moon probe had discovered this mysterious base on the Moon.

Another strange photo from a YouTube video on secret bases on the Moon.

The 2012 spoof sci-fi movie *Iron Sky* featured a plot where the Nazis escaped to the Moon at the end of WWII and built a gigantic Moon base in the design of a swastika.

The Nazi Moon base from the 2012 movie *Iron Sky*.

The Moon base featured in the television show *Space: 1999*.

Astronauts return to their Moon base in this European Space Agency illustration.

Two concept drawings of a Moon base that would be made with huge 3-D printers flown to the Moon by the European Space Agency.

A Russian concept for tubes to connect places on the Moon.

A New Front in a Secret War

communication.

That was probably bad enough. But there was also a second leaked memo issued the same day that almost certainly sealed Kennedy's fate.

The second memo issued November 12th is a classified memo to the director of the CIA, who Kennedy may or may not have known, was "MJ-1," the secret head of the MJ-12 organization. The memo orders the director of the CIA to provide NASA with files on "the high threat cases" with an eye toward identifying the differences between what Kennedy calls "bona fide" UFOs and

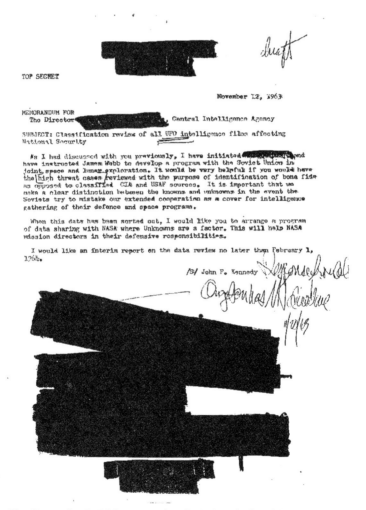

The Kennedy "UFO" memo, dated 10 days before his murder.

Hidden Agenda

any classified United States aircraft. He informs the CIA director that he has already instructed Webb to begin the cooperative program with the Soviets (which confirms the other, authenticated memo) and that he would then like NASA to be fully briefed on the "unknowns" so that they can presumably help with sharing this information with the Russians.

Now, what these memos, taken together, show is that the program of cooperation between the United States and the Soviet Union was real, and that Kennedy intended to share our most highly classified technology and UFO secrets with our arch Cold War enemy. We don't know for certain that this was the last straw, but when you read lines like, "As you must know Lancer has made some inquiries regarding our activities which we cannot allow," it takes on a much more chilling meaning. What we do know for sure that only 10 days after issuing these two memos, Kennedy was dead. The only questions are who killed him, and why?

The circumstances of John F. Kennedy's death at the hands of one or more assassins in Dealey Plaza in Dallas, Texas on November 22 1963 have been hotly debated. But what can no longer be debated is an undeniable "NASA connection" to the events of that day.

On his trip to Texas, the night before the assassination Kennedy was celebrated at a dinner in Houston hosted by Congressman Albert Thomas. Thomas was a close friend of Vice President Lyndon Johnson and one of the most powerful men in Congress. More importantly, he was in charge of all the budgets and funding for NASA, and he vehemently opposed Kennedy's plan to form a joint Moon program with the Russians. The next day, Thomas flew aboard Air Force One with President Kennedy on the short flight to Dallas. In what seems to be a bizarre occult ritual, First Lady Jacqueline Kennedy was given a bouquet of red roses upon her arrival in Dallas, instead of the traditional "yellow rose of Texas."

Red roses are a very bad portent in Texas, where the red rose signifies blood, or violent death. It's as if they were telegraphing that John Kennedy wasn't leaving Texas alive. Jackie Kennedy even commented on the strangeness of getting red roses instead of yellow when she was handed them. From there, the motorcade

A New Front in a Secret War

proceeded through the streets of downtown Dallas, and to its inevitable brush with fate in Dealey Plaza.

Now, there are a million theories about what happened in Dealey Plaza that day, some of them sensible, some of them crazy. I cover many of them in my books *Dark Mission* and *Ancient Aliens and Secret Societies*. But the bottom line is this: if you can prove more than one gunman shooting at the president, you have a conspiracy. A lone nut gunman could still be a lone nut gunman. But two shooters, that's a conspiracy that leads to all kinds of new questions. So the question must be asked, is there any real evidence of a second shooter that fateful day in Dealey Plaza?

There are piles of evidence that suggest that Lee Harvey Oswald was in the 6th floor window of the Texas School Book Depository, and that he fired at President Kennedy at least twice. What most people don't know is that there's also a lot of evidence for a shooter on the infamous "grassy knoll."

Despite multiple witnesses' testimony that a shot or shots came from the grassy knoll area, these claims have been largely dismissed over the years. But in the late 1980s, a new witness named Gordon Arnold came forward to claim he had been on the grassy knoll that day, and that shots rang by his head.

Arnold stated that he had been on the grassy knoll, right in front of the picket fence, when a shot rang past his left ear. Having just completed basic training for the Army, Arnold reacted instinctively and dove to the ground. A few moments later, he was

Gordon Arnold on the grassy noll where he heard a shot ring past his left ear.

Hidden Agenda

confronted by a man in a Dallas police uniform, but not wearing a hat. He pointed his high powered rifle at Arnold and demanded the film in his camera. Arnold gave it to him and the man disappeared behind the fence. Shaken, Arnold said nothing for almost 25 years about what he had seen that day.

So the question becomes, is there any evidence that Arnold was on the grassy knoll that day? Well, for one thing, Senator Clayton Yarborough, who was in the motorcade behind the president's car, stated in an interview that when the shots rang out, he saw a uniformed man hit the deck, just as Arnold said he did that day. But perhaps more important is the content of a photograph of the grassy knoll area taken just moments before the president is killed by a shot to the head. It is known as the "Mary Mormon photograph."

Mary Mormon is a woman who was standing on the grass opposite the grassy knoll and took a photograph just an instant before the president was hit with the fatal head shot. Most people don't know this, but the shot from the rear by Oswald that passed through the president's neck would probably have killed him anyway, since it severed the nerve that tells the heart to beat and the lungs to breathe. But there's no question that the head shot killed the president instantly.

In the early 1990s the Mary Mormon photograph was subjected to detailed analysis for the first time in support of an

The infamous "Badgeman" photo enhancement.

A New Front in a Secret War

A&E mini-series titled "The Men Who Killed Kennedy." Under scrutiny, the photo yielded solid evidence supporting Gordon Arnold's story.

Under photographic enhancement the photo shows not only Gordon Arnold in his Army summer uniform holding a camera in front of his face, but right next to him is a figure nicknamed "Badgeman." You can see that Badgeman is wearing a Dallas policeman's uniform. The patches and the badge are in exactly the right place, and his arms and elbows appear to be extended as if he's holding a rifle. Right where you'd expect to see the muzzle blast is a bright flash of light.

Observers also note that the figure not only has clearly defined facial features, but like Arnold stated, he is not wearing a police hat.

So here we have indisputable visual evidence of a gunman on the grassy knoll, spotted in a photograph and corroborated by eyewitness testimony which pre-dates the photographic work. But did he fire the fatal head shot, or did Oswald?

Everybody knows about the Zapruder film, but again, on the other side of the street there was a woman talking film of the president the moment he was shot. Her name was Marie Muchmore, and her color film is really amazing. If you slow it down frame by frame, you can actually see the microsecond when the president is struck in the head, *from the front*.

When you look at this frame capture, you can see a spray of

Frame capture from the "Marie Muchmore" film of the Kennedy assassination.

Hidden Agenda

red mist just in front of the president's forehead. That's actually a spray of blood from the bullet which has just entered his skull. You can also see that Kennedy's hair is being pushed straight up from the wake turbulence of the rifle shell, but the back of his head and hair are undisturbed. This is clear proof that the bullet entered from the front, not the back, as the Warren Commission always claimed. What you're actually seeing is the split- econd between life and death for President Kennedy. And you're seeing absolute proof that he was struck in the head by a shot from the front—from the grassy knoll.

Historical records are clear on what happened next, After fleeing the scene, Oswald was spotted a few miles away and was confronted by a Dallas police officer named J.D. Tippit. According to witnesses, Oswald fired several shots at Tippit, killing him, and then fled and tried to hide in a nearby movie theater, where he was captured. After being held overnight, Oswald was killed in the basement of the Dallas police headquarters by Jack Ruby during what was supposed to be a routine transfer to a more secure jail.

The weird thing about Ruby is that the Warren Commission claimed he was not in Dealey Plaza at the time of the assassination, but photographs show him standing outside the Dallas police headquarters in the Plaza just minutes after the assassination. The morning he killed Oswald, he somehow managed to get by layers and layers of security to get to Oswald, and to this day no one knows how he got into that basement with a gun.

Following Tippit and Oswald's deaths, the only remaining member of the team of assassins would have been Jack Ruby, who never talked before his death in prison in 1968. There is no question that Ruby was somehow involved, because his presence in the basement of the police headquarters is otherwise completely unexplainable. Somebody had to let him in. Whatever Ruby's involvement, there is no question who the death of President Kennedy benefitted.

What a lot of people don't know is that the week that President Kennedy was away in Texas, the Oval Office had been scheduled for a remodeling. The very tasteful blue-gray carpet in the office was torn up and replaced with a garish, blood red carpet that seems completely out of character with Kennedy's

A New Front in a Secret War

tastes and preferences. It's almost as if whoever was in charge of the remodeling knew Kennedy wasn't coming back from that trip to Dallas. The person in charge, it turns out, was Vice President Johnson's wife, Ladybird. When now President Johnson arrived back in Washington, he returned to the remodeled, red carpeted Oval Office.

What a lot of people don't know is that Lyndon Johnson was a 33rd degree Scottish Rite Freemason, and as such he was very aware of the trappings of office and the power of symbolism. The red carpet is symbolic of the blood of the King, Kennedy, whom he had killed. He even used Kennedy's old rocking chair. The symbolism is Johnson sitting on the throne of his predecessor, while symbolically swimming in his blood. The whole thing is actually ghoulish.

With Kennedy out of the way, his initiatives were quickly swept aside. Johnson and Congressman Albert Thomas moved quickly to stop the space cooperation with the Soviets, and the silver certificates were withdrawn from circulation immediately. The dark forces behind the public space program were now free to go to the Moon and retrieve any alien technology they found there, without concern that Kennedy would expose it to the public or share it with the Russians.

Hidden Agenda

Chapter 5
The Parallel Secret Space Program

As we have seen, before he was assassinated in Dallas, President Kennedy had engaged in a very aggressive program to explore and exploit the new frontier of space. As we can see from "The President and the Press" speech railing against the secret societies and the burned memo warning of Kennedy's inquiries into the activities of MJ-12, Kennedy was a real thorn in the side of the secret space program groups. As we look back on Apollo and the other programs he initiated in his May 1961 "We Choose to Go to the Moon" speech, it seems to me that Kennedy was very frustrated by his inability to find out what the CIA, MJ-12 and NSA were up to. In fact, it is my conclusion that he decided early on to create a second, parallel and secret space program wrapped in the mantle of the public front presented by NASA.

The evidence for this is quite strong. Much of it you can read about in my books *Ancient Aliens on the Moon*, *Ancient Aliens and Secret Societies*, and *Dark Mission*. It bears a closer examination and update here. The first place to look for evidence of a hidden agenda within NASA is in the formation and personnel of the agency itself.

NASA, as we know it today, actually evolved from several diffuse earlier organizations. One, called the National Advisory Committee for Aeronautics, or NACA, was the primary source of early NASA personnel. The NACA director, Dr. Vannevar Bush, was influential in many major aerospace projects and companies and was a founding member of MJ-12, according to the Eisenhower briefing memo. He was also co-founder of Raytheon (still a major defense contractor) and was director of the Office of Scientific Research and Development, which oversaw the Manhattan Project. He was also President Roosevelt's scientific advisor and played a key role in bringing many of the German/Nazi rocket scientists,

Hidden Agenda

like Wernher von Braun, to the United States.

Shortly after the formation of NASA back in 1958, then-President Eisenhower surprised many in the scientific community by passing over the highly respected, apolitical Hugh L. Dryden (Director of NACA since 1949, after Dr. Bush left) and naming T. Keith Glennan instead as NASA Administrator.

Glennan had been president of the Case Institute of Technology in Cleveland, was a former member of the Atomic Energy Commission (with the highest level security clearances), and was a staunch Republican to boot. Glennan also had an extensive military background, having served as Director of the U.S. Navy's Underwater Sound Laboratories during World War II. As the first Administrator of NASA, Glennan immediately set up a compartmentalized internal structure, far more akin to an intelligence gathering agency than a civilian scientific program. Dryden was appointed to the post of Deputy Administrator. Under this structure, Glennan would furnish the administrative leadership (and policy direction) for the new entity, while Dryden would function as NASA's scientific and technical overseer.

On a single day in 1958, NASA absorbed more than 8,000 employees and an appropriation of over $100 million from NACA when it was first formed. Under the terms of the Space Act, accompanying White House directives and later agreements with the Defense Department, the fledgling agency also acquired the Vanguard rocket project from the Naval Research Laboratory, and the Explorer project and other space activities at the Army Ballistic Missile Agency (but not the von Braun rocket group). It also obtained the services of the Jet Propulsion Laboratory, hitherto an Army contractor, and an Air Force study contract with North American for "a million-pound-thrust engine," plus other Air Force miscellaneous rocket engine projects and instrumented satellite studies. In addition, NASA received $117 million in appropriations for "military space ventures" from the Defense Department. Von Braun was appointed director of the new Marshall Space Flight Center in July 1960 and given the task of developing the rockets for the new agency.

Glennan moved quickly to establish the Agency's monopoly over space exploration. One of NASA's first acts was to commission

The Parallel Secret Space Program

the Brookings Report, which makes crystal clear the discovery of extraterrestrial artifacts fall under the dark blanket of "national security."

I do not believe this was an accident. What Brookings essentially did was give NASA political cover for what its real mission was all along—the retrieval of alien technology from the surface of the Moon for reverse engineering purposes. Stonewalled in his efforts to gain access to the anti-gravity technology of the secret space program, Kennedy decided to do an end run via NASA, and then develop his own secret space fleet.

When he took office in 1961, Kennedy moved swiftly to replace Glennan and restructure NASA to accomplish the major goal of his Presidency: placing a man on the Moon by 1970. To this end, Kennedy (on the specific recommendation of Vice President Johnson) appointed James E. Webb as the new Administrator. It is under Webb, a 33° Scottish Rite Freemason—that the true influence of the various secret societies within the new agency came into its own.

But that's another book…

Within a few months, Webb had appointed many fellow

Kennedy meets James Webb for the first time in the Oval Office.
Note his uncomfortable body language.

Hidden Agenda

Freemasons to positions of high authority within NASA. Kenneth S. Kleinknecht was appointed as director of Project Mercury. Kleinknecht was the brother of C. Fred Kleinknecht, who was the Sovereign Grand Commander of the Supreme Council, 33°, Ancient and Accepted Scottish Rite Freemasons, Southern Jurisdiction for the United States of America, from 1985 to 2003. Their father, C. Fred Kleinknecht, Sr., was also a 33° Scottish Rite Freemason and member of the Scottish Rite Supreme Council.

"Kenny" Kleinknecht had already been selected in 1959 as one of two "single points-of-contact" between NASA and the DOD. In this dual role, he was able to monitor information that traveled back and forth between Project Mercury and the Pentagon. With a lengthy history as an engineer in a variety of black military programs in the 1950s, he was ideally suited for this job. He went on to become a technical assistant to Mercury Program Director Robert Gilruth in 1960, and became Project Manager for Mercury on January 15, 1962. Kleinknecht also became Deputy Project Manager for the Gemini Program, and was the Apollo Program Manager for the Command and Service Modules.

If there was a plan for the Masons to place "their" men at the highest levels of the space program, it could not have been more successful.

Given his distaste for "secret societies, secret oaths and secret proceedings," it is surprising to me that Kennedy allowed this, or that he allowed von Braun to appoint Nazi war criminals like Kurt Debus to run the space center at Cape Canaveral. Either he was forced to accept these appointments as a practical matter simply to get the job done, or he planned to use the secret societies to his advantage. Perhaps he was following a "keep your friends close but your enemies closer" working philosophy. As we have seen, this was a very dangerous game fraught with risks to life and limb.

In any event, Kennedy and the newly minted space agency moved quickly to get the Moon missions going. Proposals of all types were welcomed at NASA, some of them quite bizarre and at least one of them very secretive.

At the end of July 1961, just a few weeks after Kennedy's announcement, Lockheed Aircraft Corporation executive E.

The Parallel Secret Space Program

J. Daniels met with Paul Purser, Technical Assistant to Robert Gilruth of NASA, to discuss a possible study contract for a very wild idea. Lockheed's proposal was to send a spacecraft with a single astronaut on a one-way trip to the Moon.

With the political objective simply to beat the Russians to the Moon, Lockheed proposed that the astronaut be deliberately stranded on the lunar surface. He would then be resupplied by rockets essentially shot at him (presumably for several years) until the space agency developed the capability to bring him back. Purser, in a move worthy of Pontius Pilate, kicked Daniels upstairs and referred him to NASA Headquarters.

As bizarre as the idea was, it might have worked save for the navigational problems of getting to the Moon. However, once von Braun had worked those out in 1963, there's no reason the idea couldn't have worked.

NASA histories state:

> Almost a year later, in June 1962, John N. Cord and Leonard M. Seale, two engineers from Bell Aerosystems, urged in a paper presented at an Institute of Aerospace Sciences meeting in Los Angeles that the United States adopt this technique for getting a man on the moon in a hurry. While he waited for NASA to find a way to bring him back, they said, the astronaut could perform valuable scientific work. Cord and Seale, in a classic understatement, acknowledged that this would be a very hazardous mission, but they argued that 'it would be cheaper, faster, and perhaps the only way to beat Russia.' There is no evidence that Apollo planners ever took this idea seriously." [1]

Project Horizon

While the Lockheed proposal seems extreme and very risky, there was a far more ambitious and very "2001-ish" proposal made by the Army very early on that is epic in its scope and technical objectives: Project Horizon.

On June 8, 1959, a group at the Army Ballistic Missile Agency (ABMA) produced a report for the U.S. Department of

Hidden Agenda

the Army on moon base project entitled "Project Horizon, A U.S. Army Study for the Establishment of a Lunar Military Outpost." The proposal was wildly ambitious, and called not just for the exploration of the Moon with a few spacecraft and astronauts, but for establishing a full-blown moon base staffed by up to 20 astronauts by 1967. The recently declassified project proposal states the military and scientific objectives and requirements for the Project as such:

> The lunar outpost is required to develop and protect potential United States interests on the moon; to develop techniques in moon-based surveillance of the earth and space, in communications relay, and in operations on the surface of the moon; to serve as a base for exploration of the moon, for further exploration into space and for military operations on the moon if required; and to support scientific investigations on the moon.

The permanent outpost was predicted to cost between $6 billion and $7.5 billion dollars (a fortune at that time, over $48 billion in today's dollars) and was to become operational in December 1966 with 12 soldiers inhabiting the base. The plans called for an incredible number of Saturn rocket launches into low

Artist's concept of Project Horizon lunar base.

The Parallel Secret Space Program

Earth orbit where the various components would be assembled into a space station for transport to and eventual placement on the lunar surface, where the empty tanks would be converted to a well-shielded, habitable base. Various moon crawlers and construction vehicles were to be transported and assembled in situ by the astronauts, and used to expand the base as needed.

The launch schedule called for 40 Saturn launches in 1964 and the first delivery to the Moon by January of 1965. By April of 1965, Project Horizon called for the first manned landing by two astronaut-soldiers, who would begin the build-up and construction phase immediately. By November of 1966 the base was to be completed and staffed by a "task force" of no less than 12 men, preferably more. To support this effort, the project called for a total

Scientists differ on whether sites should be underground in a lunar crater or "ocean" or if they should be blasted out of the sides of mountains.
—*Martin*

A lunar construction vehicle working on the Moon. Artist's conception from the declassified Project Horizon proposal.

Hidden Agenda

of 61 Saturn I and 88 Saturn II launches up to November 1966. By the time the base was completed, at least 245 tons of material and equipment would have been transferred to the Moon.[2] Then from December 1966 through 1967, the first operational year of the base, 64 more launches were scheduled which would transfer an additional 120 tons of equipment to maintain and expand the base.[3]

The actual design and layout of the base was to be an "L" shaped configuration, consisting of buried or partially buried cylindrical metal tanks approximately 10 feet in diameter and 20 feet in length. Two nuclear reactors would be transported to the Moon and located in pits to provide shielding and power for the operation of the living quarters and for the construction of the permanent facility. Empty cargo and propellant containers would be assembled and used for storage of bulk supplies, weapons, and life support equipment like oxygen and water.

Two types of surface vehicles were to be used, one for lifting and digging, and another for extended distance trips necessary for hauling, reconnaissance and rescue. The base was to be serviced by a "direct ascent" landing and return vehicle which could carry up to 16 astronauts at a time and return them safely to Earth. A lightweight parabolic antenna erected near the main quarters would provide communications with Earth. At the conclusion of the construction phase, the original construction camp quarters would be converted to a bio-science and physics-science laboratory.

The proposal was taken seriously enough that Wernher von Braun, then the head of ABMA, appointed Heinz-Hermann Koelle (another Nazi) to head the project team at Redstone Arsenal in Alabama.[4]

One of the first tasks of Koelle's team was to consider a site for the new Army base. Because of the limitations of the rockets then in development, possible sites for the base were restricted to the central equatorial disk of the Moon, from 20 degrees north and south to 20 degrees west and east.

Three locations very near to each other were selected as the most ideal: the northern part of Sinus Aestuum, near the crater Eratosthenes; the southern part of Sinus Aestuum near the mysterious Sinus Medii; and the southwest "coast" of Mare

The Parallel Secret Space Program

Imbrium, just north of the Montes Apenninus Mountains.

But where the proposal really gets weird is when it begins to mention weaponry. The plan calls for an "intermediate station" in orbit around the Earth for use as a scientific laboratory and weapons platform. Specifically, the documents mention "Instrumentation for test of earth-to-space weapon effects." But the document doesn't specifically mention the Soviet Union as the adversary they would be using these Earth-based "space weapons" against. So who are they planning on fighting?

The document also makes it clear that from the Army's perspective, there can be no exploration of space without the military exploitation of space:

> The military potentialities of space technology, which the United States would prefer to see channeled to peaceful purposes, are greater than general public discussion to date suggests. Military space capabilities are technically inseparable from peaceful capabilities which are well worth pursuing in their own right. Reconnaissance for merchant-ship land patrol and for peaceful mapping of resources can also be used to locate military targets. Communications to improve global relations can also be used to control military forces. Rockets for cargo and passenger delivery can also carry thermonuclear weapons.
>
> Satellites designed to return men from an orbit to a preselected point can also deliver bombs.

It's clear from the documents that Horizon would have been a military base first and a scientific/research lab second. With that in mind, provisions for ordinance were obviously a key consideration. An overland attack by Soviet forces was considered a possibility, and one proposal was that conventional Claymore mines be modified specifically to puncture pressure suits of any soldiers from the invading Soviet hordes. There was also an even more bizarre accommodation that was proposed for the bases defense: nuclear weapons.[5]

The Project Horizon proposal suggested that the base could

Hidden Agenda

A "Davey Crockett" nuclear rocket.

be armed with up to a dozen unguided low-yield nuclear missiles. These weapons were to have been modified to operate in the harsh lunar environment, and were to be fired from a 155mm recoilless rifle tripod. They could deliver a 0.1 kiloton nuclear warhead about 2.5 miles away.

Obviously, even transporting such weapons to the Moon would be a highly dangerous endeavor, one fraught with scientific and political risks that make it seem hardly worth the effort. If the rocket carrying the weapons went off course and crashed near a Soviet base, there would be a risk of them falling into Russian hands and being used against the U.S.'s own base. Even worse, if the rocket crashed *into* the Soviet base and the weapons detonated, the risk would be all-out war with the Soviets. So why take the risk?

Many in the UFO community have suggested that the nukes were not for defense against the Soviets, as this would be far too risky (as we have just discussed) and provocative. They suggest that these weapons were designed not to protect the base from Russian attack, but rather to fend off attacks or interference from hostile *alien* interlopers on the Moon. While I find this idea plausible, I did not read anything in the Project Horizon documents that led

The Parallel Secret Space Program

me to believe there was a secret, alien agenda behind the base. While provisions for defense against hostile ETs would certainly be a consideration, I find no evidence that it was the primary driver behind the proposal.

That said, there is no doubt that the Horizon documents represent a detailed blueprint for constructing an operating lunar base in short order. I see no reason why these plans couldn't have been carried out behind the scenes, in parallel with the public NASA space program. There have always been constant rumors about secret U.S. bases on the Moon, and here we have the evidence of the germination of that idea. It would have been a fairly simple thing to implement this plan over the next few decades, and staff it with Gary McKinnon's "non-terrestrial officers."

My suspicion and speculation is, that is exactly what they did.

If the Project Horizon documents were somewhat unproductive in developing a paper trail on the parallel, secret space program that I believe Kennedy created to counter MJ-12, new evidence has emerged about an old mystery that shows there was such a hidden agenda buried deep within the Apollo program itself.

Apollo 12

I've written before about *Apollo 12* on a couple of occasions, in my books *Ancient Aliens on the Moon* and *Dark Mission*. Exactly what happened on the lunar surface in the Ocean of Storms in November, 1969 has always fascinated me for one simple reason: We never saw any of it.

After the roaring success of *Apollo 11*—which as I pointed out in my previous books as well as in *Ancient Aliens and Secret Societies* was a purely symbolic mission—*Apollo 12* was the most critical of the rest of the missions because it was the first to get down to the real, secret business of Apollo: salvaging abandoned alien technology for reverse engineering.

Apollo 11 had landed basically in the middle of nowhere, a desolate plain in the Sea of Tranquility, purely to perform a short moonwalk and so that Buzz Aldrin could conduct a sacred Masonic ceremony in the Lunar Module (LM)—the consecration of a Masonic temple on the Moon. As I have discussed in my previous books, Aldrin took a ceremonial Masonic apron to the Moon with

Hidden Agenda

him and admittedly performed a "communion" ceremony (actually a Masonic rite) in the LM before he and Neil Armstrong's historic first moonwalk. Once that task was accomplished and the men had returned safely to the Earth, NASA's attention turned to the next voyage to the Moon, *Apollo 12*.

Astronauts Pete Conrad and Al Bean were to be the third and fourth men to walk on the surface of the Moon, and after the disappointing and blurry images from *Apollo 11*, we were all looking forward to watching them for several hours on two Extravehicular Activities (EVAs) on the much improved television cameras that *Apollo 12* carried.

Even though a color video camera had been carried to the Moon on both Apollo 10 and *Apollo 11*, the astronauts were not allowed to take the high-resolution camera into the LM or to the lunar surface. *Apollo 12* was the first mission to allow the camera to be taken to the surface and it was specially modified for the task. However, about 42 minutes into telecasting the first EVA, astronaut Alan Bean "inadvertently" pointed the camera directly at the Sun while preparing to mount it on the tripod. According to the Apollo 12 Mission Report, the Sun's extreme brightness burned out the video pickup tube, rendering the camera useless.[6] The report did not mention that this was something Bean had been specifically trained *not* to do.

I have always been suspicious of this "accident." I just don't see how a highly trained, professional astronaut like Bean could have made such a basic mistake as pointing the camera right into the sun and burning out its optics. The end result is that the world got to see almost nothing of what Bean and Conrad did on the surface of the Moon, since there were no provisions for a backup TV camera aboard the LM. To me, this is especially suspicious in light of just what I think *Apollo 12*'s real mission was, and where it had landed.

Thirty-three months before *Apollo 12* landed on the Moon (yes that number is Masonic, and no I don't think it is a coincidence) *Surveyor III* landed in essentially the exact same location on April 20th, 1967, which just happens to be Adolf Hitler's birthday (*Seig Heil!*). The ostensible reason for *Apollo 12* to land near *Surveyor III* was to demonstrate that a precise landing point could be selected

The Parallel Secret Space Program

Jagged, crystalline spires poking miles above the surface of the Moon.

and executed by the Apollo team. The real reason is somewhat different.

As I've argued in *Dark Mission* and *Ancient Aliens on the Moon*, I believe the Moon, especially the front side, is mostly covered by towering crystalline, glass-like structures which acted as a makeshift meteor shield for the various alien basses operating on the surface below. In a lunar environment, glass would be about as strong as steel and obviously far more lightweight and amenable to construction. These "Crystal Towers of the Moon" reach staggering heights, in some cases as much as 20 miles above the surface. The visual evidence for their existence is covered in my other books, so I won't go into detail on that here. Suffice it to say I am more than satisfied I can prove their existence to anyone with a reasonably open mind.

The point is, these crystal towers were there, and NASA knew it. That's why they landed *Apollo 11* far away from them, and why LM pilot Buzz Aldrin activated both the landing *and* side-mounted docking radars during the *Apollo 11* descent, leading to the famous "1202" alarm incident which almost scuttled the first manned lunar landing.

That also, in my opinion, is why the color camera was not taken to the lunar surface on *Apollo 11*. Even though they were pretty sure they were well away from any large-scale visible

Hidden Agenda

ruins, NASA wanted to be sure nothing untoward appeared on the world's television screens. So they ordered Armstrong and Aldrin to leave the color video camera in the Command Module while they took only the crappy black and white low-res video camera to the surface. They even made sure that the pictures would be made worse by applying a "high-pass" filter to the video transmission, effectively reducing the detail visible in the TV signal.

Besides its proximity to the *Surveyor III* landing site, *Apollo 12*'s landing site is also bizarre and unique in that it is only about 122 miles from the *Apollo 14* landing site, which was the next successful lunar landing and exploration. Scientifically, it makes no sense to have two of the six Apollo landings take place so close together. These were rare opportunities, after all. Geologically, the areas would have been almost identical, and the scientists would have learned very little from *Apollo 14* they didn't already learn from *Apollo 12*. Unless of course "geologic science" wasn't the objective of the two landings at all.

Comparison of photographs and films from the two

Frame capture from NASA film "Pinpoint for Science" showing towering glass structures over the horizon from the *Apollo 12* landing site.

The Parallel Secret Space Program

neighboring landing sites shows the same distant, over-the-horizon semi-transparent structures visible from both sites. It is fairly obvious that investigation and reconnaissance of these ruins was the primary and secret prime directive of both missions. After having landed and completed the symbolic and political objectives of *Apollo 11*, NASA and the secret civilian space program decided to "go for it" on *Apollo 12* and take the risk of landing amongst the ruins. That's why NASA landed *Surveyor III* in the same location three years earlier.

By safely landing *Surveyor III* at the site, not only did NASA prove that it was possible to navigate through the ruins and set down relatively near them, they provided a proven, safe flight path through the towering glass structures that Apollo 12 could follow. Thirty-three months later, that's exactly what they did.

But, it is what they did immediately *after* they landed that has now caught my intense interest. While there is almost no video of Conrad and Bean's activities on the lunar surface during their mission, there is a great deal of photographic material which, under enhancement, reveals that they were literally down among the ruins the entire time, doing who knows what. My guess is, looking for alien artifacts to return to Earth and back-engineer.

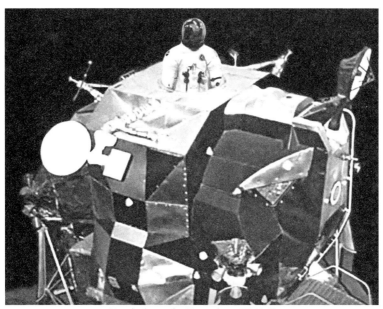

Depiction of a "stand-up EVA."

Hidden Agenda

But what they could not have known, especially from the *Apollo 11* experience, which used a vastly inferior camera, is exactly what we the viewers at home would be able to see on our TV screens. So they did something very sneaky and very secret: they performed an unrecorded, and unacknowledged, stand-up EVA.

[Chap5N-7]
Depiction of a "stand-up EVA."

In the space parlance of the day, a stand-up EVA is where one of the astronauts stands on the ascent stage engine cover in the Lunar Module and looks around outside the main hatch at the top of the cabin. There was one such publically acknowledged stand-up EVA performed on the *Apollo 15* mission, but until recently, there had never been a publically acknowledged stand-up EVA performed on *Apollo 12*. In a sense, there still hasn't.

Until a report issued by a NASA historian in 2006 entitled "NASA/TP-2006-213726—The Apollo Experience Lessons Learned for Constellation Lunar Dust Management," no one ever knew there was a stand-up EVA on *Apollo 12*. But there, on page 1 of the document, the truth slipped out for the first time:

> *The blowing dust caused by the Apollo 12 LM landing appears to have been worse than that of Apollo 11. In fact, a standup extravehicular activity (EVA) was performed by the crew to assess the site prior to performing lunar surface EVAs because blowing dust completely obscured the view during landing.*[7]

So there it is, right there in black and white. I don't think this particular historian had a clue what a bombshell she had just dropped in her obscure scientific study. While attempting to produce a report on the hazards and issues with lunar dust for future Moon missions, she inadvertently revealed an activity that NASA had made some effort to conceal for some decades. Specifically, that Conrad and Bean had performed a stand-up EVA just after landing.

The Parallel Secret Space Program

This is in stark contrast with the official record, which not only shows that there was no S-EVA on *Apollo 12*, but that there were only two official EVAs on that mission, neither of which involved standing on the ascent stage casing and looking out the roof of the spacecraft.

Because of dust and various other factors, during descent Conrad had a difficult time determining exactly where he was in relation to the target object, *Surveyor III*, and the "Surveyor crater" in which it rested. This led to him landing with the LMs "back" to the *Surveyor* spacecraft and with the area in front of them saturated with bright sunlight, it would have been very difficult to visually determine exactly where his Lunar Module *Intrepid* was in comparison to where it was supposed to be. These two factors alone would have been sufficient to justify a S-EVA. There's just one problem: Why keep it a secret if dust and location were the only reasons to pop the hood and have a look around?

A YouTube videographer operating under the handle "LunaCognita" has done an excellent job deconstructing the timeline from the official NASA records.[8] There was a gap of 4 hours 49 minutes after landing and before the first public, official EVA moonwalk where the S-EVA had to have taken place. We know it took place in this period because the dust study says so (see above). Around 37 minutes after landing, the astronauts were supposed to be taking astronomical observations and entering the data into an onboard computer to help nail down their exact position. According to the public radio transcripts, at that time Conrad claims that he entered an incorrect digit into the platform alignment program computer, and then tells Houston he'll have to do the entire data entry over again from the top. NASA agrees and then Conrad says "Bye-bye," indicating he's going to go silent for a while. This then becomes the cover story for the clandestine stand-up EVA that was obviously planned all along.

It's important to note there is no record of this second data entry ever happening. The official transcript shows multiple comm breaks over the next several minutes. The first one is quite long, listed officially in the transcripts as 9 minutes and 3 seconds of dead air. Applying a stop watch to the actual audio recordings shows that it is actually 12 minutes and 38 seconds of dead air. What

Hidden Agenda

exactly NASA and the astronauts did with the extra 3 minutes and 30 seconds is a mystery. Later, there are two more comm breaks totaling over 8 minutes and 30 seconds.

At that point, just before the clandestine S-EVA begins, CAPCOM (Capsule Communications) officer Jerry Carr tells Conrad that the computer alignment is done, and "the computer is yours." Conrad replies with a simple "Thank you," followed by an 18 minute 45 second period of absolute radio silence from *Intrepid* on the lunar surface. Communications resume at Mission Elapsed Time 111:58:43.

There are a couple of further points to note here. The LunaCognita guy seems to buy into the idea that the stand-up EVA was solely to help determine the location of the spacecraft after landing. I believe this could be part of the reason, but as I

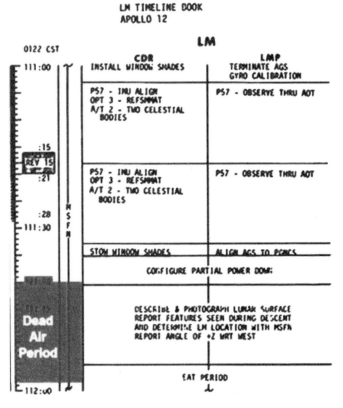

Excerpt from *Apollo 12* LM timeline book indicating the period of the "19-minute gap."

stated earlier, why would such an off-schedule activity have to be done in secrecy? Further, there is the P 57 platform alignment program of the computer itself. The astronauts used these devices specifically to get a fix on their position, and would not need to get up and look around to determine their position. Planetary and stellar alignments would give them their exact location without this exercise.

In any event, the nearly 19-minute gap in the public transmission from *Intrepid* would provide plenty of time for both of the astronauts to step onto the ascent stage casing, stick their heads out and relay their observations back to mission control in Houston. They would also have had the opportunity to take full 360-degree photographic panoramas of the area around the landing site, which they almost certainly would have done, although those photographs have never been revealed.

According to the *Apollo 12* lunar timeline book, during this almost 19-minute period of radio silence, the astronauts were supposed to be looking out the windows and describing what they were seeing over the public radio channel. In fact, they did not do so until well after the period when the secret S-EVA took place. Instead, fully 1 hour and 10 minutes after landing on the Moon, no description had been given of the terrain seen out the windows. Also, there is nothing in the transcript to indicate they had taken off helmets or pressure suits, which they would have needed to perform a S-EVA. Normal procedure would have been to at least take off their gloves and helmets. If however, they had planned a clandestine S-EVA which required the cabin to be depressurized all along, they would have left their suits and helmets on and left themselves plugged into the LM's environmental control systems.

As to that long period of dead air, there are in fact several hints that this 19-minute gap was no gap at all. Contrary to what most people believe about the Apollo missions, there were in fact several different ways in which the astronauts could secretly communicate with NASA back on Earth. One way is to use the bio-medical telemetry feed, which had duplex capability and could be used for private voice communication.

There is in fact confirmation from *Apollo 17* astronaut Gene Cernan that these private radio channels existed. In his

Hidden Agenda

autobiography *The Last Man on the Moon*, Cernan makes two references to these secret channels of communication with the engineers and scientists back on Earth.⁹ "Stripped down to our liquid cooled underwear, we had a quick dinner, debriefed with the guys on Earth *via a private radio loop*, and played with some of the rocks we had stowed in the cabin boxes," he writes.

Later in the book, Cernan write about using a private comm loop again. Amid concerns that the Palestinian terror group Black September had threatened *Apollo 17* and the crew's families, Cernan describes a private conversation with NASA Director of Flight Crew Operations Donald K. "Deke" Slayton over the covert radio channel: "After we had had some chow and settled down, Deke told me on a private radio loop that everything was fine at home. 'I talked to Barbara [Cernan's wife] and everything is okay,' he said. Not a peep from Black September." Neither of these conversations appears anywhere in the official NASA public records or transcripts from *Apollo 12*. So there is no doubt that private, classified and off-the-record conversations took place on all the Apollo missions between the astronauts and NASA.

But, after the 19-minute gap ends, we get additional confirmation that a stand-up EVA took place. According to the official transcript, the next transmission between *Intrepid* and Houston is with an entirely new CAPCOM officer, Dr. Edward Gibson. Sometime between the time the public channel went silent and when it picked back up nearly 19 minutes later, astronaut Jerry Carr had been replaced as CAPCOM by Gibson. This is especially significant because Gibson was not just one of the rotating CAPCOM astronauts. For *Apollo 12* he was the specially designated Extra Vehicular Activity CAPCOM. In other words, he was the only NASA official allowed to communicate to the astronauts during their EVAs. The *only* reason Gibson would have been on the radio with Conrad and Bean after the 19-minute gap is if they were in fact performing—you guessed it—an EVA!

The mystery gets even more interesting when you study what is actually said between the various parties involved. "Well done Intrepid," is the first thing Gibson says when they switch back to the public channel. What, exactly was "well done?" After 19 minutes of silence? For all Gibson knew, at least according to

The Parallel Secret Space Program

the public record, Conrad and Bean could have been masturbating for the previous 20-odd minutes. He was clearly congratulating them on their successful, and completely clandestine, S-EVA. Also, Conrad never acknowledges the CAPCOM change, which are always noted by Houston and the crews. This is because the changeover from Carr to the EVA CAPCOM, Dr. Gibson, had already taken place at the beginning of the secret S-EVA over the private channel.

The next thing Gibson says is also very telling: "Say, we're standing by with a LM consumables update." "Consumables" is, of course technical-speak for the oxygen and water that the astronauts needed to survive on the lunar surface. Bean had already reported that the H_2O and oxygen levels were "good," so why would NASA need an update on their current levels? Obviously, if they had depressurized the LM before opening the hatch for the S-EVA and then used a substantial amount of oxygen to repressurize the cabin afterwards, NASA would need to know what the new levels were. There is no other explanation.

Yet another hint comes from what happens next. Both Conrad and Bean then proceed to return to the public script they have been given and describe for the American people what they see outside their spacecraft, albeit 19 minutes late. But there are discrepancies in Bean's descriptions that further indicate a clandestine S-EVA took place.

It is well established that visibility out of the tiny, triangular front windows of the Lunar Module is very poor. In fact, astronaut Dave Scott, who performed a publically acknowledged stand-up EVA on *Apollo 15* was asked in an interview after that mission if the view is better from the top hatch than through the windows: "Oh, yeah. Oh, without a doubt," he said. "The windows are very restrictive."

Yet Bean twice in the radio recordings makes reference to a panoramic view of the landing site at a time he supposedly hadn't been outside the spacecraft. "You can see quite far in all directions," he says. Followed a bit later by: "There are blocky rimmed craters that are visible in almost every direction." How, exactly, does he know this if he has not yet been outside the LM?

To me, this all adds up to one undeniable truth. *Apollo 12*'s

Hidden Agenda

secret stand-up EVA was conducted for one true purpose, to make a determination of just how much the public would be able to see on the fancy new color camera they had brought to the Moon. The idea that the S-EVA was conducted just so they could get a fix on their position doesn't add up because they already had that information and because there would be no need to keep such an EVA secret. If the real reason was to make sure that distant glass-like ruins could not be seen on TV, that would certainly be sufficient reason to conduct the S-EVA in the dark, away from the public's prying eyes.

And obviously, they must have ultimately feared what the public would see because just under an hour later, Bean, in virtually his first act on the lunar surface, deliberately pointed the camera at the sun and burned out the optics.

Honestly, I really doubt this was what JFK had in mind.

(Endnotes)

1 Paul E. Purser to Gilruth, "Log for week of July 31, 1961," 10 Aug. 1961; House Committee on Science and Astronautics, Astronautical and Aeronautical Events of 1962: Report, 88th Cong., 1st sess., 12 June 1963, p. 112; "Apollo Chronology," MSC Fact Sheet 96, n.d., p. 19; John M. Cord and Leonard M. Scale, "The One-Way Manned Space Mission," Aerospace Engineering 21, no. 12 (1962) : 60-61, 94-102.

2 http://www.history.army.mil/faq/horizon/Horizon_V2.pdf

3 http://www.history.army.mil/faq/horizon/Horizon_V1.pdf

4 https://en.wikipedia.org/wiki/Project_Horizon

5 http://www.ancient-code.com/declassified-documents-reveal-project-horizon-the-lunar-outpost-of-the-us-army/

6 https://www.hq.nasa.gov/office/pao/History/alsj/a12/A12_MissionReport.pdf

7 "NASA/TP-2006-213726 - The Apollo Experience Lessons Learned for Constellation Lunar Dust Management." Sandra A. Wagner, Johnson Space Center.

8 https://youtu.be/kXkXCBiAHQ4

9 *The Last Man on the Moon: Astronaut Eugene Cernan and America's Race in Space* by Eugene Cernan (Author), Donald A. Davis (Author), St. Martin's Griffin, ISBN-13: 978-0312263515

Chapter 6
The Alien Reproduction Vehicle

Image of the Alien Reproduction Vehicle taken in Provo, Utah in 1967.

For decades, rumors have persisted of a dark conspiracy to hide the existence of a secret American space program. Evidence has shown that in the mid-1950s, major American defense contractors backed by the U.S. government were frantically searching for breakthroughs in anti-gravity technologies which would have made such a program possible. But direct video evidence and verifiable eyewitness testimony have always remained elusive...

Until now.

In the last 20 years, a great deal of visual evidence and personal accounts have revealed proof that the U.S. has not only continued their clandestine anti-gravity research, they appear to have perfected it. If these new lines of evidence turn out to be true, then not only have the American people and the world been lied to for decades, but secretive elements inside the U.S. government may have exploited these technologies for their own unknown—and perhaps nefarious—purposes. As far back as the late 1940s, rumors have persisted of a secret government program to reverse engineer and exploit technologies obtained from otherworldly

Hidden Agenda

sources.

Documents show that all the way back to the Cape Girardeau saucer crash in 1942, the U.S. government was trying to understand and exploit alien technologies obtained from the crashed saucers. Specifically, the documents mention that fiber optics were first discovered on the Cape Girardeau craft, and then developed over the next 30 years by the U.S. military/industrial security state.

By the time of the Roswell crash five years after the Cape Girardeau incident, the U.S. government appeared to have a plan in place to deal with the new "alien problem" that was becoming an increasing public relations issue. The MJ-12 group, hastily formed after the Roswell incident, came up with a plan to distribute alien technology to American industry.

Whether the Roswell incident was a crash of a German-derived flying wing test aircraft or a genuine flying saucer from another world is debatable. But if the Cape Girardeau incident was real, then the alien technology from that craft could account for the amazing leaps in technology by U.S. industry in the last 70 years.

The story of the Alien Reproduction Vehicle, or ARV, is

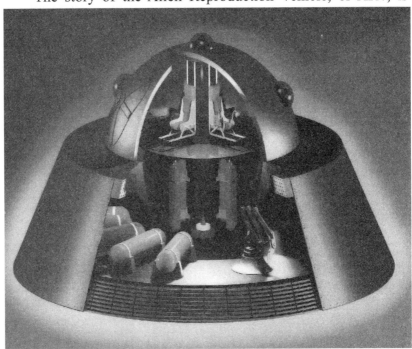

Cutaway view of the ARV as described by Brad Sorenson.

The Alien Reproduction Vehicle

a pretty interesting example. Mark McCandlish, a production illustrator for the aerospace industry, was planning to go to an air show in California with a buddy named Brad Sorenson. At the last minute, a job came up and McCandlish couldn't go. Sorenson went without him to the event at Norton Air Force Base in November 1988. At the event, Sorenson found himself mixed in with a group of VIPs who were taken to a special hangar on the far side of the base. The area was blocked off from the general public with huge black curtains. Inside the VIP area, Sorenson saw a long-rumored secret aircraft called the Aurora, which was a special high altitude pulse-jet aircraft that was Top Secret at the time. He also saw something even more amazing.

According to Sorenson, next to the Aurora were three "flying saucer" type vehicles, lined up in a row and floating above the hangar floor. He was told they were under development by Lockheed, and were designated the "Alien Reproduction Vehicle," or ARV.

Sorenson said the saucers were completely silent, but were definitely operational. One of them had several side panels removed. He was allowed to look inside of one the ARVs, and he described a number of features of the interior to McCandlish. They included details like a series of storage tanks laid out on the lower deck like spokes of a wheel. It was unclear what they contained (N/B gas?), but he was under the impression they were crucial to the vehicle's operation. He also observed two pilot's seats on the upper deck, of the proper size for a fully grown human being to use.

McCandlish later drew up the craft from Sorenson's observations, and concluded that the craft was assembled cross-sectionally in something like 40 segments, which he estimated weighed five tons each. For something that massive to be hovering above the hangar floor silently there had to be some kind of anti-gravity effect going on. Sorenson and the other VIPs were also shown a video of the same ARVs performing test maneuvers. He said that the craft were shown hovering in groups, then darting back and forth and even shooting straight up into the sky to high altitude at incredible speeds, all without making a sound. The video also said that the Skunk Works team's nickname for the

Hidden Agenda

The ARV and Schauberger's Repulsine side by side.

ARV was the "Flux Liner."

My observation is that the ARV as described by Sorenson bears more than a passing resemblance to Viktor Schauberger's Repulsine, and I think it's possible that they had a common evolutionary path. This is exactly how aerospace development works. Every commercial plane in the sky is really just an incremental development of the 707, which was the prototype for the modern jetliner. The bigger question is, what was it doing in a public air show if it's still a classified project?

To this day, no one knows why Sorenson was allowed to see what he saw that day. It may have been simply that the Air Force or Lockheed wanted to push the technology into the public arena for financial reasons.

The fact that it appeared in the air show at all indicates one of two things: Either this was a mistaken distribution of the classified version of a secret Lockheed project, or the program isn't classified anymore! If the program has been declassified, it means that this is a deliberate "leak" telling people that they can now go find patents and other information in archives and research papers. It could be a hint that the anti-gravity propulsion systems that made the ARV work are now kind of an "open source" technology.

This could lead to widespread development of advanced propulsion systems and weapons. Such an approach would probably accelerate development of the technology in the ARV, but it would necessarily come at great risk as well. If this is a deliberate leak aimed at inspiring outside groups or inventors to investigate and further develop the technology, then it could violate the key precepts of the Bookings doctrine. It would be an effective acknowledgement that extraterrestrials not only exist, but have been visiting Earth for quite a while.

The Alien Reproduction Vehicle

Illustration of the Air Force's Project 1794.

But McCandlish and the story of the ARV are not immune from criticism or conjecture that places the craft in an alternate light. It's been argued by some of the usual suspects in the debunking crowd that the Sorenson/McCandlish story might have a perfectly logical and non-ET explanation. Critics of Sorenson and McCandlish have argued that what Sorenson saw in the hangar that day was nothing more than a misidentified jet aircraft called "Project 1794."

Project 1794 was an Air Force attempt to create a jet powered flying disk. It was given to AVRO Aircraft Limited of Canada in the late 1950s for prototype development. It was based on something called the Coanda effect, which basically uses high pressure thrust over a disk shape to achieve lift.

It's pretty obvious that the inspiration for Project 1794 came from UFO and flying saucer reports. The Army and Air Force were trying to duplicate the performance of craft that had been reported in various sightings. Their first attempt at a prototype was the Avro Air Car, which was tested in 1958.

It's pretty clear from the declassified Project 1794 documents that the Air Force had high hopes for the project. The initial test

Hidden Agenda

bed was a gas turbine powered 18-foot diameter two-seater called the VZ-9 air vehicle. They tested it extensively, but found that it couldn't get very far off the ground or generate much in the way of directional thrust. After a few months of testing, they went back to the drawing board and came up with some modifications, but they didn't perform any better.

It's really clear from reading the Project 1794 documents that the Air Force was very intent on building its own flying saucer. There was going to be a larger, 25-foot diameter full scale production aircraft called the Y-2, which would carry a single pilot and vertically take off and land. There was even a further development called Project Silver Bug that was supposed to be 108 feet in diameter!

Unfortunately, the prototype model underperformed pretty dramatically, and the Air Force lost interest. After a few more years of experimentation, the Avrocar was retired and the program was cancelled in 1961. Today the prototype is in the Smithsonian National Air and Space Museum. But these inconvenient facts didn't stop critics and debunkers from accusing Sorenson and McCandlish from simply misidentifying Project 1794 aircraft as alien flying saucers.

What the critics said is that the Project 1794 documents, which were declassified in 2012, "proved" that Sorenson was simply mistaking conventional technology for alien technology. They argued that there were many similarities to the Project 1794 proposals. They cited things like the centrally located pilot's cockpit, the central compressor accounting for "heavy copper plates" mentioned by Sorenson, and they claimed he mistook the radially mounted jet engines for storage tanks.

Really, their claims are kind of absurd. For one thing, no prototype matching Sorenson's description was ever built, and the Project 1794 illustrations don't match very well either. Just because they're both disk shaped doesn't mean they're the same. Plus, the whole project was cancelled in 1961, 27 years before Sorenson saw the ARV.

The whole thing gets even more ridiculous when you consider that Sorenson testified that the ARVs he saw were hovering in place, and that they made no sound. Obviously a jet

The Alien Reproduction Vehicle

engine pumping out enough thrust to keep the disk hovering inside a hangar would be deafening. There's just no way it can be the same craft, or a further development of a conventionally powered aircraft. Also, for the debunkers' theory to fit, there must be some indication in a document somewhere that the Avrocar project had either not been cancelled or had changed names. There are no such documents that anyone can find. From all indications, the Air Force's attempts to build a jet powered flying saucer ended in 1961.

But if the Avrocar doesn't explain what Sorenson saw, what does? Is it possible that the ARV has been developed privately and actually used in a quiet war in space? Not only is it possible that the ARV exists, we have actual footage of the spacecraft being tested in low Earth orbit.

There are two very impressive sets of footage from NASA space shuttle missions STS-48 and STS-80. Not only do they show that something very much like the ARV exists, but the vehicles seem to be deployed in what looks like some kind of secret war going on right over our heads.

The only real questions are, who are we fighting, and what are we fighting for?

Since the beginning of the space shuttle program, NASA cameras have captured all manner of strange and unusual sights in the space around the shuttle and the International Space Station. Footage of bizarre, firefly-like objects have been observed, along with objects that move rapidly past the camera's range of view, and even small objects that seem to suddenly change direction and increase in size and brightness.

Without fail, NASA apologists and UFO debunkers dismiss these sightings as misinterpreted ice crystals or space junk floating past the camera lens. But over the years, a few very exceptional pieces of footage have surfaced that seem to show something far more interesting and far harder to explain going on in the hard vacuum of space above our planet. They seem to show powered space vehicles performing impossible maneuvers right before the camera's eye.

On September 15, 1991, the space shuttle *Discovery* on mission STS-48 was cruising along in low Earth orbit at about

Hidden Agenda

17,500 miles per hour. The camera in the crew cabin was pointed toward the tail end of the shuttle with the equipment bay doors opened, showing the Earth below and the space above. An independent researcher named Don Ratsch was in the habit of taping the live feed from NASA Select TV, and he recorded some incredible footage that night.

As you watch the video live feed, there are three major events that take place over the course of a few hours. Event #1 occurs with the shuttle over Western Europe. One by one, a series of five lights travel radially from behind the tail section. Then an extremely bright blinking light appears, traveling up from behind the left engine sheath. This bright light hovers in the middle of the arc composed of the five smaller lights. Then the camera inexplicably pans down into the cargo bay, focusing on a piece of equipment. After a minute, the ground controller requests that she be given control of the camera. After a short pause, the astronaut on the radio complies. The camera then pans back up into the tail area, and the lights that were there just moments before are now gone.

A few hours later, the main event, or Event #2, takes place. *Discovery* is orbiting 355 miles above Rangoon, Myanmar. You can see the city lights below as well as a lightning storm taking place in the upper atmosphere. There are at least five unexplained lights hovering in the shot between the shuttle and the Earth below and the space above which absolutely are not stars. They pretty much keep stationary with the shuttle, and never disappear behind the horizon as stars or planets would.

The female CAPCOM comments to the astronauts that while they were "looking at Orion" the troops in mission control were getting a fabulous light show from the electrical storm. As the footage continues, an object suddenly appears at the horizon, right in the middle of the frame. It traces along the Earth's horizon for a few seconds, moving right to left, and then a bright flash of light goes off in the upper left corner of the screen. The object in the middle of the frame makes an immediate 45-degree angle turn, accelerates through the airglow layer of the atmosphere, and disappears into space. As it does so, two streaks appear from the Earth below, one aimed at the main object and the other aimed

The Alien Reproduction Vehicle

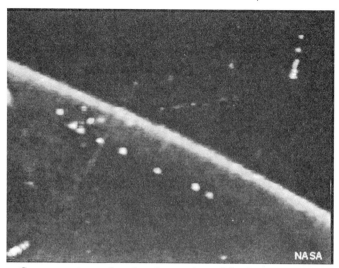
Screen captures showing the motion of the STS-48 UFO.

at another of the hovering objects. These projectiles, or whatever they are, both miss their intended targets.

It's actually not too surprising that both of these shots missed their target. The calculated speed of the main object was Mach 73, or 54,000 miles per hour! When it turns, it accelerates to Mach 285, or 211,000 miles per hour! There is nothing we have ever flown that can match even the slower speed, much less survive the G-forces of making the turn.

To make a 45-degree angle turn like that and accelerate to that speed, which this object does in 2.2 seconds, the G-forces would be something on the order of 14,000 Gs. That's the equivalent of dropping a 10-story building on top of you! Obviously, we don't have anything in the public domain that can make a maneuver like that, and even if we did, the pilot would be smeared all over the inside of the spacecraft. He couldn't survive.

The only way corporeal tissue could survive that kind of acceleration is if the craft had some kind of inertial dampening field around it. This kind of a protective field is implied in the work of T. Townsend Brown and even the Nazi Haunebu saucers. Within the bubble of the field, the relative speed and G-forces are negated.

When the STS-48 footage first surfaced, it quickly made the rounds to various UFO conferences and was hotly debated on

Hidden Agenda

TV shows like CNN's *Larry King Live*. The critics were quick to attack it.

NASA and the NASA apologists went on the offensive immediately after the STS-48 footage was leaked. The first thing they did was shut down the live transmission feed on shuttle missions and insert a delay timer so they could cut off the transmission if anything like this happened again. Then when citizens began to ask for high resolution versions of the film, they found there was a 14-minute gap in the provided footage where some more incidents happened a bit later on.

NASA followed that up by sending out a hack spokesman named James Oberg to various TV shows to attack the footage and claim it was just ice crystals. They claimed that the flash was just a maneuvering thruster firing.

The NASA argument basically came down to this: All the objects in the field of view are simply ice crystals close to the spacecraft from a urine dump on a previous orbit. They claim the flash is a small maneuvering thruster called a vernier rocket, the action of which pushes the "ice crystal" in a different direction.

The problem with this explanation is that some of the objects are actually hovering, and don't change direction abruptly as they would if they were hit by a thruster blast. So the argument came down to trying to find a way to prove one way or another whether the objects were close by or far away. After a lot of debate about that, NASA critic Richard C. Hoagland did an analysis that all but proved that the target object was at the horizon more than 1,700 statute miles away.

Hoagland's video on the subject was brilliant because he pointed out that the main object, the one that makes the 45-degree turn, is partially obscured by the horizon when it pops into view. What's actually happening is that the craft is in the Earth's atmosphere and comes over the horizon and into the shuttle's field of view. That means that the object has to be at least as big as the shuttle to even be visible from that distance.

Another way to tell that the object is at the horizon is that when the flash goes off, it's actually glowing very brightly. When it makes the sudden turn, it then passes through the airglow layer and into space, where it fades out. This is exactly what we would

The Alien Reproduction Vehicle

expect to see from a craft utilizing the Biefeld-Brown effect for propulsion. The brightness is caused by the electrostatic charge of the field around the spacecraft interacting with the atmosphere. These highly charged particles create the glowing effect through a kind of static discharge. When it crosses the airglow limb and moves into space, there's nothing for the charged particles to "push" against so the glow dissipates. That's exactly what we'd expect to see if it was a powered vehicle at the horizon, and it is in fact exactly what we do see.

If it was really an ice crystal, as NASA flatly argued, it would be moving into the daylight because the shuttle was very close to the day/night terminator at the time. That means the object would have to get brighter, not dimmer.

NASA's explanation also can't account for the streaks that go by the main object and another craft in the lower right of the frame. They don't even mention them. It turns out that the velocity of the streaks were calculated to exactly match a newly developed Star Wars weapons system called "Brilliant Pebbles."

Brilliant Pebbles was originally called "smart rocks," and it was a kinetic anti-missile weapon fired from an orbital space platform. It fired watermelon-sized chunks of steel at super high velocities to intercept ballistic missiles. An electromagnetic accelerator called a Railgun was used to shoot the projectiles at the target. On the STS-48 footage we see these projectiles, which

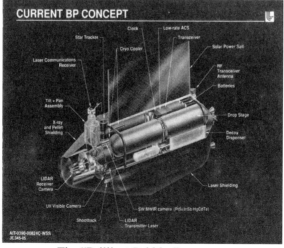

The "Brilliant Pebbles" concept.

Hidden Agenda

exactly match the performance envelope of Brilliant Pebbles, shoot across the screen at two of the objects. They miss by a mile, but that's exactly what you'd expect from something that has the capabilities that an ARV possesses. Something like the Alien Reproduction Vehicle could easily dodge a weapon as sluggish as Brilliant Pebbles. It's like trying to catch a road runner with a turtle.

However, these counterarguments didn't deter the critics of the STS-48 footage. Notable lackey James Oberg, a former NASA employee and political operative, came up with another argument to counter Hoagland's analysis. Oberg argued that the shuttle was actually casting a shadow over the center of the camera frame, and he insisted that the target object did not pop over the horizon, but just coincidentally came up into the light of the coming dawn at exactly that location. The object only became visible when it moved up out of the shuttle's shadow just after sunrise. Since the video was taken near sunrise, the shuttle's shadow was pointing back nearly parallel to Earth's horizon, and the ground was still dark. This would require that it be close to the shuttle. The proximity to the horizon line would be coincidental.

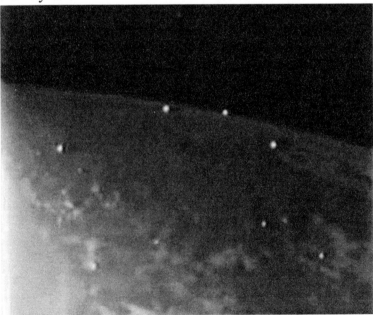

Hovering, glowing objects from the STS-48 video.
Where's the "shuttle shadow"?

The Alien Reproduction Vehicle

The shadow argument is completely destroyed by the fact that another bright object crosses the screen from right to left, passing nearly the same point at which the "target object" first appears, and remains visible. Other objects also glow and hover in the center of the screen, right where Oberg argues this mythical "shadow" should be cast. This is proof that no such "shuttle shadow" exists. This reinforces Hoagland's claim that the object rises from behind the physical horizon, since there is no counter mechanism to explain a close-by object suddenly becoming visible.

Oberg argues that Hoagland's statement that the object should brighten as it "rises into the sunlight" is also false because no brightening takes place after the object is in full sunlight. Once in full sunlight, no further brightening occurs. But he does not address the significant and indisputable *dimming* of the target object just after it crosses the airglow boundary. According to Oberg's own reasoning, this should not occur. The target object starts *below* the airglow boundary, increases in brightness as it crosses the boundary, and then dims noticeably and continues dimming. It should resume its previous level of luminosity and stay constant until it is out of frame, according to Oberg's "small object, close to the shuttle" model. Since it does not, optical effects *cannot* account for the brightness increase as the target object crosses the boundary. Hoagland's model is the only one which matches the bright/brighter/dim/dimmer behavior of the target object. It should also be noted that Oberg does not mention the two "projectiles" which cross in front of the camera. They also significantly dim after crossing the airglow layer. Oberg also fails to mention that the projectiles come from *below* the shuttle, almost 90 degrees from the direction of the target object's post-flash flight path. Obviously, a blast from a thruster could not account for two such disparate flight paths.

But if that weren't enough, Oberg's analysis was also criticized by University of Nebraska physicist and astronomer Dr. Jack Kasher, who did a frame-by-frame breakdown of the STS-48 footage.

Kasher's frame-by-frame analysis has shown that the target object begins to *accelerate* some 1.2 seconds after the flashes, making the minimum distance to an alleged "ice particle" nearly

Hidden Agenda

two miles. At this distance, such an ice particle would have to be enormous to be visible, and the blast would not account for the observed acceleration due to thrust dissipation. Further, NASA telemetry shows that all the vernier thrusters fired at least twice during the duration of the tape and no similar flashes were visible in any of those instances.

So if the thrusters can't account for the change of direction by the main target object or the two projectiles shooting at it, what does? Kasher has the answers to that too.

By the time Event #2 takes place, the shuttle *Discovery* was actually over Western Australia, and the "shots" taken at the space objects are traceable back to Exmouth Bay near the North West Cape military facility, and the Pine Gap military facility in central Australia, respectively. This means that both projectiles had a military origin, and are most likely examples of anti-ballistic missile technology.

Another key piece of proof that we are looking at spacecraft under intelligent control is that the main target object doesn't just make a maneuver to avoid the shot, it actually *hesitates* before it does so. The object is stationary for almost half a second, an impossible feat for a supposedly tumbling ice particle. The simple fact that the target object comes to a complete stop for half

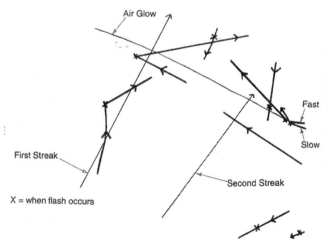

Dr. Jack Kasher's diagram of the STS-48 objects' flight paths.
Used with permission.

The Alien Reproduction Vehicle

a second is proof that it is under powered flight and intelligent control. Oberg has never even tried to explain that.

But the stunning events of STS-48 in 1991 would pale in comparison to an even more bizarre series of events that took place a few years later on another shuttle mission: STS-80. If STS-48 is the smoking gun of UFO footage, then STS-80 is the gunman on the grassy knoll of UFO footage.

STS-80 was a much later mission of the space shuttle *Columbia* in December of 1996. Late in the evening on December 1st, the shuttle's aft bay cameras were focused in on city lights over Santiago, Chile, at about 33 degrees latitude. As the shuttle cameras are focused in on the city below, suddenly this huge glowing object just shoots out of the middle of downtown and into space so fast the cameras can't even track it! Nicknamed "the tadpole" because of its odd, swimming motion, this object is so big and so fast that it can't possibly be any kind of conventional aircraft.

Still shot from the STS-80 footage showing "the tadpole" streaking into space from the outskirts of Santiago, Chile on December 1,

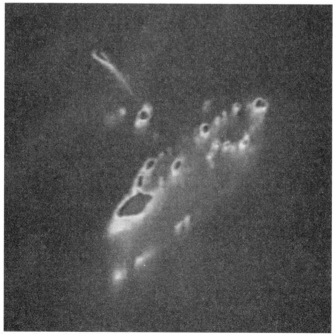

Still shot from the STS-80 footage showing "the tadpole" streaking into space from the outskirts of Santiago, Chile on December 1, 1996.

Hidden Agenda

Video experts have noted that the object doesn't streak across the frame, it emerges from another bright area on the ground. It also seems to "wiggle" on its flight path, which no natural object could possibly do. It really looks like some kind of giant tadpole shooting into space! After this event, the camera, which is being remotely controlled from Mission Control, pulls back to get a wider field of view. We see stars slowly setting behind the horizon. Then when it zooms back up again, things start to get really weird.

We then get a couple minutes of thunderstorm activity. But about 3 minutes 37 seconds into the video, the star of the show suddenly appears! It comes in from the lower right quadrant of the picture and passes *in front* of the horizon! This means that it is not a star, but is actually somewhere between the shuttle and the clouds below. Its velocity relative to the space shuttle is considerably higher than the 17,000 miles per hour the shuttle is travelling. It flares up brightly as it hits the atmosphere, and then begins to slow down, as if the atmospheric drag is not only electrostatically charging the air around it, but creating resistance and slowing it down. There is absolutely no way that this object can be an ice crystal, because if it was, travelling at that speed, it would have instantly burned up in the atmosphere. Instead, it starts to pace the shuttle. It brightens incredibly as it hits the atmosphere, and then almost immediately begins to slow, as if someone just slammed on

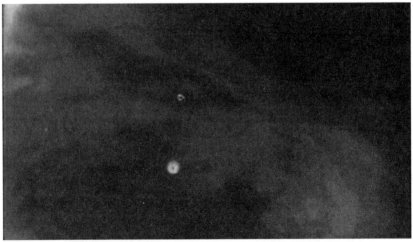

STS-80 target object flaring up as it hits the atmosphere several hundred miles below the space shuttle *Columbia*.

The Alien Reproduction Vehicle

the brakes. It slows down until it matches the speed of the space shuttle at about 4 minutes (remember, that's over 17,000 miles per hour), and then it *station keeps* with the shuttle for over two minutes. As it holds its relative position to the shuttle, you can see other objects on curved trajectories orbiting around this thing while it just hovers there in one spot.

Then at 5 minutes in, another enormous object moving very rapidly just appears in the center of the screen, moving right to left. It *also* decelerates and matches the shuttle's speed, something no natural object could do. About 30 seconds after the second object appears, the first object changes direction and starts to retrace its course *back* the way it came! The second object remains stationary above the clouds while the first object moves back into space and forms a triangle with two other objects that have now appeared! The remote camera operator then focuses in close-up on the three objects, and keeps tracking them. As it pulls back out, the objects become too dim and the show ends.

It's pretty simple. Ice crystals and debris in space do not change direction. They do not change speeds; they do not speed up or slow down. They do not survive reentry into the Earth's atmosphere at thousands of miles per hour. They do not slow down and keep steady pace and distance from the space shuttle. And most especially, they do not hover for two minutes and then travel back in the same direction they came from!

The only explanation that fits the behavior of the objects in the STS-48 and STS-80 footage is controlled performance by powered vehicles. Not just that, but vehicles that can perform far beyond the capabilities of our currently accepted aerospace technology and physical sciences. In other words, they're flying saucers. To my mind, what we have to be looking at is some kind of weapons system test on our own spacecraft. You are literally looking at the Alien Reproduction Vehicle in action!

A lot of people have suggested that this is a Star Wars weapons test against alien space craft. But does that really make sense? The cameras were poised to capture these events on video. Did the ETs phone NASA and tell them they would be at such and such a place and time, so have the cameras ready?

No, I don't buy that. The only scenario that fits is that NASA

Hidden Agenda

was used to record a test firing of our own weapons systems against our own ARVs.

But that raises an even more disturbing question. If we are developing our own space weapons systems and testing them against UFO level technology, who are we building these systems to fight against?

Chapter 7
The Whistleblowers

Over the years, a number of credible witnesses have come forth to openly discuss their experiences with the secret space program. Foremost among them are the likes of Lieutenant Colonel Philip Corso, former NATO Command Sergeant Major Robert O. Dean, Sergeant Clifford Stone, who was part of the Army's crash retrieval teams, and physicist Bob Lazar. Interestingly, while they all tell a similar story of the U.S. government's involvement with extraterrestrial races, they take very different views on the possible dangers posed by such contact.

According to one whistleblower, MJ-12 designated him as the primary point of contact on the release of alien technology for reverse engineering purposes.

Philip J. Corso was a United States Army Lieutenant Colonel who had served four years on the National Security Council in the Eisenhower White House in the 1950s. During his time in the military, he was placed in charge of the Foreign Technology Desk under Lieutenant General Arthur Trudeau, who headed Army Research and Development. In this position, he would take technological artifacts obtained from Russian, German and other foreign sources, and have private American companies reverse engineer that technology. In 1997, he published his only book, a memoir entitled *The Day After Roswell*, in which he revealed just how "foreign" some of these sources actually were.

The real bombshell in *The Day After Roswell* was when Corso admitted that some of the technology he passed on to American defense contractors, like the transistor, came from alien sources. He said he was given these pieces of technology without any attribution as to where they had come from, but it was generally accepted that the source was extraterrestrial. Certainly fiber optics, lasers, integrated circuits and transistors were far beyond the industrial capacity of the United States at that time,

Hidden Agenda

Army Lieutenant Colonel Philip J. Corso.

and perhaps even the Germans. So where had they come from?

Even though Corso was never told the specific origin of the technology he was given to distribute, he speculated that the source was the Roswell crash, which he had been briefed on while he served in the Eisenhower administration.

According to Corso, he gave the transistor to Bell Labs in New Jersey. Bell had been working on developing a transistor circuit, but had made very little progress and had reached a dead end. The integrated circuit boards retrieved from Roswell and Cape Girardeau were unlike anything even conceived of in those days, and with help from Wernher von Braun, the transistor was reverse engineered and in the public domain by the 1950s. According to Corso, the transistor age was probably accelerated by at least 10 years by giving this alien technology to the Bell

The Project Serpo Human-Alien Exchange Program

engineers to study. It is very similar to the *Terminator 2* scenario. They may not have understood everything they were studying, but it would have given them ideas and pushed them in directions they had never thought of.

The dawn of the transistor age was a virtual technological revolution. With transistors replacing vacuum tubes, radios, computers, and even spacecraft guidance systems could be reduced in size and made far more efficient. The secret was that the signals could be sent through the system without having to excite the electrons with heat in order to conduct electrical current. That means you can build switching systems at room temperature. It may have been the key to unlocking the technology of the aliens. Despite the internal consistency of Corso's testimony, many, including those in the UFO community, questioned his story.

After Corso's book came out in 1997, it was subjected to a lot of unfair scrutiny in the UFO community. People questioned his credentials, which were subsequently validated, and argued that the alleged alien technology wasn't advanced enough to be credible.

The criticism has been made that the idea of reverse engineering itself is too far-fetched. The comparison has been made that you could send a laptop back to Roman times, and they'd never figure out what it was or how it worked, much less

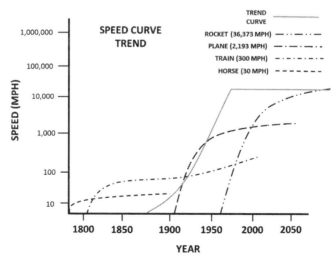

Speed trend curve show dramatic cessation of the growth of speed of transportation technologies.

Hidden Agenda

be able to duplicate it. But that assumes that alien technology is thousands or at least hundreds of years ahead of us.

But what if they aren't? What if the discovery of anti-gravity propulsion and fiber optics are a kind of technological glass ceiling? Once you break through that and can create propulsion systems that exceed the speed of light at very low energy costs, then you've reached a major plateau in your development as a technological species. If reverse engineering is impossible, why did the Army have a department specifically assigned to handle the task? And is it really so far-fetched to believe that the secret government made a major anti-gravity breakthrough in the late 1950s, when everything disappeared from the published literature?

One possible answer may be found in a simple mathematical graph showing the growth of propulsion technology over the last two millennia. If you put all of the methods of transportation on a graph, starting with the speed of marching foot soldiers and then go forward to rockets, you get a very interesting trend. Not only does the speed of transportation increase dramatically, the speed of the propulsion breakthroughs goes through the roof in the 20th century. For almost 2,000 years, the fastest things on the planet were carts hauled by beasts of burden. Then in very rapid succession you go from the horse and buggy to the steam train, the automobile, the propeller airplane, the jet engine, and then by 1960, you have the chemical rocket.

But then suddenly, it cuts off and flatlines. If you believe what the military guys are telling us, there hasn't been a single breakthrough in propulsion in almost 60 years after the graph goes almost straight up for the previous 100 years. That's basically impossible. History has shown us that technological curves like this always trend upwards, and continue to do so. Moore's Law, the observation/prediction that microprocessor speeds will double every two years—has held steady for decades. And even if it slows, it does not simply stop and have all technological improvements cease for over 50 years, as we have seen after the rocket. There should have been another major propulsion breakthrough that took us to the next level—the flying saucer level—decades ago.

So if you look at this graph, the odds are that there *had* to be a breakthrough of some kind in the last 50 years. But if that's true,

The Project Serpo Human-Alien Exchange Program

The author with Command Sergeant Major ret. Robert O. Dean.

where is it? The answer is that it's been hidden.

But perhaps, hidden in plain sight.

Robert O. Dean was a U.S. Army intelligence officer assigned to the NATO office in Brussels. He served in Intelligence Field Operations and was stationed at Supreme Headquarters Allied Powers Europe (SHAPE), the military arm of NATO. While in the Army, he achieved the rank of Command Sergeant Major, which is the highest rank a non-commissioned officer can attain. He also had a clearance level of Cosmic Top Secret, which is one of the highest possible classifications.

One day in 1964 while he was on duty at the Supreme Headquarters Operations Center (SHOC) monitoring Soviet troop movements, an Air Force Colonel, who was on duty that night as the SHOC controller, pulled a document out of the secure vault and threw it on Dean's desk and said, "Read this. This'll wake you up!"

The document, which was enormous, was titled "The Assessment" and contained the entire history of human interactions with aliens from other worlds. It was funded by NATO, took two and a half years to write and had been put together by military representatives of every NATO nation; it also included contributions from some of our planet's greatest scientific minds.

Hidden Agenda

Only 15 copies were ever published.

According to Dean, the document contained reports of civilian and military encounters with UFOs and their occupants. Some of these reports contained interviews with people who claimed to have been taken aboard UFOs.

One appendix, entitled "Autopsies," contained photographs of a 30-meter disc that had crashed in Timmendorfer, Germany in 1961. According to the report, the British army reached the crash site first and established a perimeter. Since the craft had crashed in very soft soil, it hadn't been destroyed.

Inside the craft were 12 small Grey alien bodies, all dead. There were pictures of the bodies being laid out and then put on stretchers and loaded into jeeps. There were also autopsy photos. These autopsies said that the small grey creatures looked as if they were clones of some kind, with no alimentary track. They did not ingest, or process, food as we do, nor did they have any system for elimination. It appeared that they were incapable of reproduction. The craft was cut up into six pie-shaped pieces and flown off to Wright-Patterson Air Force Base in Ohio.

But what was most intriguing about "The Assessment" was the conclusions it reached...

"The Assessment" made five major conclusions:

- First, our planet and the human race have been studied by several different extraterrestrial civilizations. Four of these races have been identified visually. One race looked the same as the human race. Another race was about the same height, stature and structure as we are, but had a very grey, pasty skin color. The third race was the small, large-headed and small-bodied Greys. And the fourth race was reptilian, with vertical pupils and lizard-like skin.

- The second conclusion was that these alien visitations had been going on for at least 200 years, and probably longer.

- The third conclusion was that these extraterrestrial visitors did not appear hostile.

- The fourth conclusion was that flybys, buzzings and various other UFO maneuvers were designed to demonstrate their technological capabilities.

- The final conclusion was that a carefully orchestrated program of some kind seemed to be underway. It began with flybys, then landings, and eventually culminated in direct contact. Dean's own assessment of "The Assessment" is that extraterrestrial visitors are intimately involved in human affairs at every level, and that they represent no threat to the human race or planet Earth.

Bob Dean is fond of saying that "extraterrestrials aren't in the government, they *are* the government." He cites numerous individuals whom he says are genetically modified Anunnaki working here to help cleanse the karma of their involvement with the human race eons ago. He's named names of high government officials and military leaders that he says are extraterrestrials here to serve humankind. It's really a very positive vision of alien visitation. Others however, do not take such an optimistic view of the alien phenomenon.

Some whistle blowers, like former Area 51 scientist Bob Lazar, warn that the alien agenda is not an altruistic one, and they claim that the secret space program is the technological backbone of a coming interplanetary war.

Bob Lazar first appeared on the UFO scene in the early 1980s. He claimed he had been employed by the defense department at a secret facility in Nevada adjacent to Area 51. He first told his story in an *Omni Magazine* article in 1981.

According to Lazar, he had physics degrees from Caltech and M.I.T., and was employed at the Los Alamos Meson Physics Facility. There are no records from these universities in his name, but Lazar claimed that his academic record was "erased" when he went public with what he knew about the secret space program.

Lazar says that after working some time at Los Alamos, he was given a new, higher clearance, a "Top Secret-Majic" clearance, and reassigned to Area 51. Since his expertise was in propulsion, he assumed he'd be working on some kind of new propulsion system

Hidden Agenda

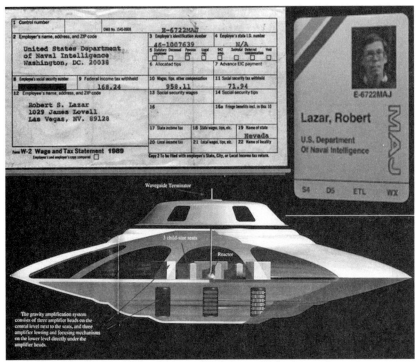

Bob Lazar's Naval Intelligence W-2 form, ID badge and a cutaway sketch of the "sport model."

there. Lazar says that a typical workday involved boarding one of the unmarked 737s flown out of Las Vegas' McCarran airport to Area 51. After arriving, the scientists and other workers deplaned and got into busses with blacked out windows for the short ride to a separate facility, which he says was called S-4. According to Lazar, S-4 had aircraft bunkers built into the hillside with camouflaged roofs to make them blend in with the surrounding terrain. When he first got there, he assumed he'd be working on captured Russian MIGs or some new experimental stealth aircraft the U.S. was developing. What he actually saw shocked him.

Lazar says that S-4 consisted of nine aircraft hangars, each of which held a different flying saucer-like craft which was being studied. He was assigned to work on a disk nicknamed the "sport model." His job was to study and reverse engineer the propulsion system. At first, he thought that what he was seeing were our own Earth-based advanced spacecraft—which some of them may have been—but when he was allowed to examine the inside of the sport model he saw tiny little seats which could only have accommodated human children... or the small aliens known as the Greys. As part of his orientation, Lazar was given a briefing on the entire history of human interaction with alien beings, sort of a 1989 version of Bob Dean's "The Assessment."

According to the briefing Lazar was given, the Greys come from the fourth planet of the binary stellar system Zeta Reticuli. The Reticulans claimed to have genetically "corrected" our evolution up to 65 times over the last ten thousand years. Divided evenly, that would be one correction every 150 years. They also referred to human beings as "containers," although containers of what wasn't totally clear.

Lazar's impression was that the term "containers" was a reference to our genetic material, but others have suggested that it may have more to do with our souls. That perhaps they view our physical bodies as simply containers for our higher selves.

According to the Greys our religions were given to us so as to, as the aliens put it, "prevent the 'containers' from destroying themselves." There were various references to religious belief

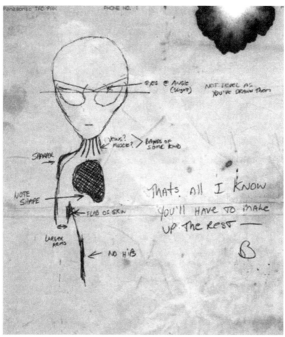

Bob Lazar's sketch of a Grey alien he saw a picture of at secret base S-4 in the Nevada desert.

systems that currently exist today. The briefing documents also claimed that at least three alien species visiting the planet were malevolent, including an offshoot of the Greys which were from Rigel. They also identified a reptilian species and a species known as the "Tarantuloids," who looked like giant spiders, as malevolent. According to Lazar, the U.S. government had made deals with the "good" Greys from Zeta Reticuli for defense technology in exchange for access to our human genome through abduction and experimentation on humans, especially human females.

Lazar also claimed that about 15 years prior to his arrival at S-4, a confrontation with the Greys had taken place there and at the Dulce base in New Mexico. It resulted in many deaths on both sides and the Greys had departed S-4 and left the humans to their own devices. He never saw an alien while he was there.

Lazar's story was highly criticized and widely dismissed, but it caught the attention of local Las Vegas reporter George Knapp.

Knapp interviewed Bob Lazar numerous times over the years and always found his story to be interesting and credible. He claims Lazar's academic record being erased seems plausible considering that's exactly the kind of thing the government would do to a whistleblower to undermine his credibility. But Knapp did do some additional investigating on his own and was able to corroborate part of Lazar's story. For one thing, Lazar has a W-2 form from the United States Department of Naval Intelligence, showing he worked for them. Knapp also found a phone book of scientists at Los Alamos for the period Lazar says he was there, and it lists him as a scientist. If he's a fraud and not really a scientist, why would he be listed as one at Los Alamos National Laboratory of all places? Clearly, he was a scientist of some kind.

If Lazar was actually a scientist, then he must have gotten his degree from somewhere. It seems like his story of his records being erased is consistent with what George Knapp found.

One of the primary criticisms of Lazar's story was how he described the system that powered the spacecraft. Lazar says that the sport model, and presumably the other craft at S-4, were powered by a tiny antimatter reactor that used an element he described as Element 115, or Ununpentium. Element 115 isn't native to our solar system, as it has a molecular density that can't be created by a yellow star. Element 115 was stable, but when bombarded with protons in tiny nuclear accelerator, it transmuted to Element 116, and released enough antimatter energy to actually amplify gravity waves and warp space.

Lazar was attacked over this by several mainstream scientists who claimed that since his ideas defied the accepted laws of physics, he must be lying. Physicist David L. Morgan says that he has scientifically refuted most of the ideas that Lazar had elaborated on in his description of the alien spacecraft, particularly its propulsion systems and use of Ununpentium, or Element 115. Morgan stated, "After reading an account by Bob Lazar of the 'physics' of his Area 51 UFO propulsion system, my conclusion is this: Mr. Lazar presents a scenario which, if it is correct, violates a whole handful of currently accepted physical theories." Morgan went on to argue that "the presentation of the scenario by Lazar is troubling from a scientific standpoint. Mr. Lazar on many occasions demonstrates an obvious lack of understanding of current physical theories."

Of course, Mr. Morgan completely misses the point. Lazar fully understands the current physical theories. He's just worked on technologies that prove that they're wrong. In 2003, Lazar's story gained substantial traction when Russian scientists succeeded in synthesizing Element 115 in a nuclear accelerator. They continue to work on stabilizing Element 115, something which Lazar claims the Grey Zeta Reticulan aliens had perfected.

Lazar's story, if it's true, indicates that the U.S. Government is pretty far along in developing a secret space fleet, apparently to defend the Earth from these alien species. The question is, what evidence do we have that this secret space fleet exists?

As we know, British hacker Gary McKinnon broke into the defense department database and found a list of "non-terrestrial officers" working on orbiting space platforms. In the mid-2000s, amateur astronomers began to find evidence that these actually existed.

An amateur astronomer who went by the name of "John Lenard Walson" (an assumed name) began posting footage of these gigantic orbiting space platforms on YouTube. He set his

Hidden Agenda

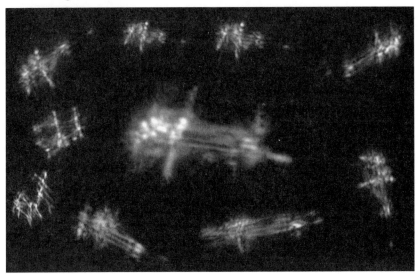

A gallery of images provided by "John Lenard Walson" of orbiting space platforms.

telescope up to track certain stars, and when he zoomed up and processed the images, he found out that these "stars" were actually huge spacecraft in Earth orbit. He was heavily criticized at the time, with some people claiming his process wouldn't work, but others have subsequently duplicated his results.

When you look at this footage, it's really incredible. You see these huge structures in orbit above the Earth, some of them looking like very advanced versions of the International Space Station. In fact, when people claimed he couldn't possibly be getting the results he showed from the equipment he had, he took footage of the ISS to prove them wrong and it matched. These giant space platforms are really creepy—very strange looking. They are so large they could easily house squadrons, if not entire fleets, of spacecraft like the Alien Reproduction Vehicle. Some of them have to be several miles across. Some of them seem to morph into other shapes, but that could just be an optical illusion. They certainly could be like giant aircraft carriers orbiting the Earth on high alert to launch interceptors. The question is, launch against which enemy?

Yet another perspective is provided by a former Army Sergeant, Clifford Stone. Stone has said the U.S. Government has tried to suppress what he had actually seen one strange day in

Pennsylvania, back in 1969.

Clifford Stone's early childhood was far different from most of ours. Growing up in the 1950s Stone, like a lot of children, had an imaginary friend. His friend went by the name "Korona," and turned out to be not so imaginary.

Stone recalls one incident where a small bird had fallen out of a nest near his home and he tried to save it. Taking the bird inside, he saw that its beak was broken and bleeding. As Clifford tried to wash the blood off its beak, the baby bird ended up drowning. When Clifford realized that he had inadvertently killed the baby bird while trying to save it, he began to cry. Korona appeared and began to question him. Why did he care so much about this bird? It was, after all, just a bird. When Stone explained that he empathized with the tiny creature, Korona simply could not comprehend that concept.

Stone says that Korona had a very real physical presence, but was invisible to other children and adults. Korona told him he was from another planet and was on Earth to study and evaluate humans. By all appearances, he was as human as you or I.

One day, Korona told him that this was not his real appearance, and asked Stone if he wanted to see what he really looked like. Young Clifford said yes, and Korona transformed into a hideous and frightening creature that terrified Stone. He ran and hid behind a couch in the family living room, but soon felt something touch his hair. Looking up, he saw the hideous face again and ran and hid behind a piano in the house. Once again Korona appeared, floating above him, and told him "you can run but you can't hide." Stone says that this point he blacked out.

Years later, Stone was drafted into the Army but was found to be medically unfit and designated 4-F. After returning to Ohio, where he grew up, Stone once again tried to enter the Army as he was very keen to serve. This time, a doctor changed his classification and allowed him to join the service, but told him to do his best to hide his medical condition.

In spite of being able to type fewer than 10 words a minute, Stone was designated a clerk typist and transferred to various bases until he ended up in Pennsylvania in 1969 with a unit that was designated to work on nuclear armed aircraft crash retrievals.

Hidden Agenda

Trained to work in high radiation environment suits, Stone's unit found themselves on a training exercise when they were suddenly called to a real crash site.

Arriving on the scene, Stone was selected from all of the members of his unit to climb a hill and look down into the crater and describe what he saw. He said that the area was glowing from fires, and when he reached the top of the hill and looked in, he saw something astonishing.

Not only was the crashed vehicle a crescent or wedge shaped craft, similar to what Kenneth Arnold had seen 22 years before, he also saw three bodies. It was clear to him that they were dead, and equally clear they were not human. Stone said that while they were certainly alien, they bore no resemblance to Korona, but were more obviously Greys.

Overcome with emotion, he begged his commanders to let him come back down the hill, but was told to continue to report on what he saw. Eventually, he was allowed to return to camp where he was debriefed by a man he called "The Colonel," but whose rank and identity he never actually discovered. The Colonel would debrief him on many occasions after that, as Stone spent the next 10 years on the Army's flying saucer crash retrieval unit.

"I was involved in situations where we actually did recoveries of crashed saucers," he said in an interview. "There were bodies that were involved with some of these crashes. Also some of these were alive," he said. "While we were doing this, we were telling the American public there was nothing to it. We were telling the world there was nothing to it."

Upset at the degree of cover-up he had witnessed, Stone began gathering documents on the U.S. military's involvement with alien species, and eventually put together reams of documents showing that ETs are well known to the U.S. government.

Stone has claimed to have catalogued 57 different species of alien life forms in the documents he has gathered. This is far different from some of the other whistleblowers, who cite only a few, but Stone says that some aliens are so similar to humans that you wouldn't even know they were aliens. "You have individuals that look very much like you and myself," he said. "They could walk among us and you wouldn't even notice the difference."

The Project Serpo Human-Alien Exchange Program

I have personally interviewed Clifford Stone and found him to be credible and his story to be consistent. What his experience indicates is that there are secret branches of government and military officials who are aware of the presence of aliens and the existence of a secret space program at all levels. Stone for one says that there are lots of government officials who would like this information released, but because of the existence of high level organizations with a great deal of power to suppress dissent, they are forced to dance around the question publically.

One of these was obviously President Reagan.

As most people know, Reagan made several speeches during his presidency in which he invoked the idea of all the nations of the Earth uniting against a common alien threat. The rumors are that under the guise of Star Wars—ballistic missile defense—the U.S. was actually launching a major initiative to defend the Earth from aliens. The implication is that after the Dulce incident, the U.S. expected the aliens to return in force to extract revenge for what had happened. The ARV, the S-4 UFOs and the Walson space platforms were all part of this new defense initiative. Supposedly, the space platforms are armed with nuclear weapons, but they're all pointed outward, away from the Earth, as if they are meant to defend against invaders, not to shoot down domestic ballistic missiles.

I don't think you can underestimate the significance of Reagan's speeches on aliens. Reagan, perhaps more than any other recent president, understood the value of symbolism. While a lot of people have commented on his UN speeches about aliens, very few have noticed that in 1992, he made just four campaign speeches supporting George Bush 41's re-election campaign. One of them was in New Mexico, a state Bush had comfortably locked up. But the location he chose was quite telling, in my opinion. The speech wasn't in Albuquerque, the state's largest city. It was in a small town known as Roswell, which is famous for only one thing; the 1947 flying saucer crash. Not only that, but he gave the speech at the old Roswell Army Airfield, right in front of the hangar where the crash debris was stored before it was sent to Wright-Patterson airfield in Ohio.

That's not a coincidence. Reagan was clearly trying to

Hidden Agenda

prepare us for what they thought was coming: an interplanetary war.

As it turns out, this was one of Reagan's last public appearances. We didn't know at the time that he was suffering from the early effects of Alzheimer's disease, and it's pretty amazing that he chose Roswell and all that that place implies as the location for one of his final speeches. He was clearly trying to send the American people a signal.

With the stage now seemingly set, the purpose and intention behind the creation of the secret space program becomes clear. Humans and aliens have had a long history of interactions, not always friendly or benign. For perhaps the last 10,000 years, alien races like the Greys have been visiting Earth, studying its inhabitants, and perhaps even intervening in our physical and emotional development. Spurred on by the threat of German aerospace technology, Soviet nuclear weapons and alien intervention, the U.S. scrambled to create a secret, highly advanced space program behind the scenes. If witnesses like Corso, Dean and Lazar are right, they have succeeded to a frightening degree. If Jon Lenard Walson's footage is real, we stand at the brink of interplanetary war, using technologies few of us understand and for an agenda we can only guess at. Unable to affect the political or military agenda, all we can do now is sit, and wait.

And watch the skies...

Chapter 8
The Project Serpo Human-Alien Exchange Program

Since the dawn of time, man has looked to the stars and asked one simple, if unanswerable question: "Am I alone?" For most of us, it seems implausible that we are the only conscious beings in all the vastness of the Universe. Other philosophers have gone farther, asking if what we know as God actually exists. And, if not God, are there other worlds like this one inhabited by other beings like us?

The answer remains frustratingly elusive to this day. But there is a mounting pile of evidence that at the very least, man's is not the only conscious, intelligent mind in the universe.

To my mind, the strongest evidence that aliens exist is that *we* exist. If, after all, the Universe creates consciousness rather than, as I believe, consciousness creates the universe, what are the chances that in all this vastness we are the only self-aware, technological beings? Given the numbers involved, I'd say precisely zero.

Almost since we first had the time to write plays and contemplate truth, we've wondered if we were alone. Are there other worlds? And what would the beings that inhabited them look like?

As far back as the late 1800s, when H.G. Wells first published his preeminent science fiction tome *The War of the Worlds*, man has considered his place in the universe. And the general consensus is—it's not a very nice place. But if the deepest fears of our collective unconscious were reflected in our novels and movies, they found validation in our earliest explorations of the solar system.

The first discovery we made when we explored space was that the Earth was surrounded by a dangerous field of radioactive energy called the Van Allen radiation belts. These belts of highly charged particles make space travel beyond Earth orbit very dangerous. The Apollo missions had to be accelerated to very high speeds to pass through the Van Allen belts quickly, to reduce radiation exposure to the astronauts.

Faced with this dangerous environment, NASA commissioned a study to help set standards for space exploration. The study considered many serious questions, including evaluating the dangers presented by space itself, and also what we might expect if we met aliens.

The first thing NASA did after its formation in 1958 was to order this cultural and scientific study from the Brookings Institution, then the most prestigious think tank in the world. The Brookings Report concluded we would probably not meet ET directly, but we might discover "artifacts" left behind on the Moon or Mars by other explorers—aliens.

It concluded that, based on the reaction to the 1939 "War of The Worlds" broadcast, society could "disintegrate" if confronted with alien ruins or artifacts. It recommended that in the event of such a discovery, NASA and the DOD should repress such evidence.

The report had a significant cultural impact. Film director Stanley Kubrick could quote from the Report chapter and verse, and he based his film *2001—A Space Odyssey* on it.

But in the mid-1980s, documents emerged that suggested that even by the time that Brookings was written, NASA and the secret government already knew that man was not alone. And moreover, that NASA would find the ruins of an ancient inhabitation of the Moon by aliens.

If the first contact with us is initiated by an alien species, to my mind, we can never make the assumption that they are here to help us. After all, it's not like we asked them for help. They must be here for their own purposes, and we are naive if we think their goals are anything other than selfish.

Whatever the motives of the aliens were at the time the MJ-12 documents were written, more than 30 years later a new and more complex view of the relationship between humans and extraterrestrials emerged in the form of another set of documents:

the Project Serpo papers.

In November of 2005, a new site labeled "Serpo.org" suddenly appeared on the internet. It contained extensive notes, recollections and declassified information from an anonymous retired military officer who claimed he had been given clearance by government officials to release the documents. The copious amounts of information purported to tell the story of interactions between the government of the United States and a race of Grey aliens from the Zeta Reticuli binary star system.

The Serpo.org data differed from a lot of other accounts of the Roswell crash and the subsequent events in several ways. Most notably, it claimed that one of the Roswell aliens had survived, that communications were established between the U.S. government and the "Ebens," as the aliens were called, and that an actual exchange program had been set up between the two worlds. Specifically, 12 humans visited the Eben homeworld, which was code named "Serpo," and two alien ambassadors traveled to Earth to be studied by humans.

The Roswell information provided claimed that there were actually two crash sites, as opposed to one. The first site was southwest of Corona, New Mexico and the second site was at Pelona Mountain, south of Datil, New Mexico. The crashes involved two extraterrestrial aircraft.

The Corona site was found a day later by an archaeology team. This team reported the crash site to the Lincoln County Sheriff's department. A deputy arrived the next day and summoned a state police officer. One live entity, nicknamed EBE, was found hiding behind a rock. The entity was given water but declined food. The entity was later transferred to Los Alamos.

The information eventually turned to Roswell Army Airfield. The Corona site was examined and all evidence was removed. The bodies were taken to Los Alamos National Laboratory because it had a cryogenic preservation system that allowed the bodies to remain frozen for research. The remains of the craft were taken to Roswell and then to Wright Field, Ohio.

According to the Serpo documents, the second site was not discovered until August 1949, when two ranchers found the wreckage. It was retrieved and the bodies and craft were sent to

Hidden Agenda

Los Alamos as well.

The live entity established communications with us and provided us with the location of his home planet. The entity remained alive until 1952. But before his death, he provided us with a full explanation of the items found inside the two spacecraft. One item was a communication device. The entity was allowed to make contact with his planet. This was later fictionalized for the film *E.T.* by Steven Spielberg.

According to the Serpo information, diplomatic relations were established and a meeting date was set for April 1964 near Alamogordo, New Mexico. One of the Eben craft was said to have gotten the landing coordinates mixed up and landed in the wrong place, resulting in the famous Lonnie Zamora close encounter of the third kind. The Ebens eventually found the base and landed and retrieved the bodies of their dead comrades. Information was exchanged. The aliens had a translation device and communication was in English.

According to the story, a diplomatic exchange program was agreed to at that first meeting. In 1965, the U.S. carefully selected 12 military personnel; ten men and two women. They were trained, vetted and removed from the military personnel database. The 12 were skilled in various specialties. Near the northern part of the Area 51 Nevada Test Site, the aliens landed and the 12 Americans left. One alien entity was left on Earth. The original plan was for our 12 people to stay 10 years and then return to Earth.

Once again, this exchange program was depicted in the final scenes of the Steven Spielberg film *Close Encounters of the Third Kind*.

But something went wrong. The 12 remained on Serpo until 1978, well beyond the 10-year planned exchange. When they were returned to the same location in Nevada, only seven of the men and one woman returned. Two had died on the aliens' home planet. Two others decided to remain on Serpo, according to the other returnees. Of the eight that returned, all have since died, the last survivor having passed away in 2002.

The story they told upon their return was quite interesting. Although the Ebens resembled the Greys, the team was told that they were not the same species, and that the Greys came from a planet in the Epsilon Eridani star system. Serpo was also uncomfortably hot, and the research team eventually had to move to the northern part of Serpo to get some relief from the heat.

The team had free run of Serpo, and they were allowed to explore as they wished. The only time they got into any trouble was when they attempted to photograph some Eben children, and were told in no uncertain terms not to do so again.

The Ebens also had a different understanding of time than we do, and the team soon found that the laws of physics as we understand them did not apply on Serpo. The Ebens told the team that physics is different in every star system, because the physical laws are dictated by the spin energy of the bodies that make up each solar system.

The Ebens told the team that Serpo was about three billion of our years old. The two suns were about five billion years old. The Eben civilization was estimated to be about 10,000 years old. They had migrated to Serpo from another planet. The original home planet of the Ebens was threatened with extreme volcanic activity, forcing the Ebens to relocate to Serpo in order to protect their civilization. This had occurred some 5,000 Earth years ago.

But this is where the story turned a bit dark. According to the Ebens, they had a great interplanetary battle with another race, which they refused to identify, about 3,000 years ago. The Ebens lost many thousands of lives in their battle, but they completely annihilated all of their enemies. The team took this as a stern warning not to underestimate the Ebens' military capabilities or their ruthlessness in war. The Ebens have never fought another war since. They've long ago been space travelers for the past 2,000 years, and they first visited Earth about that same time.

Obviously, the news that the Ebens had fought an interstellar war would have really alarmed our government. The Ebens did warn us that several other alien races within our galaxy were hostile, and that they themselves stay away from those races. The debriefing document never stated the name of their enemy, probably because they no longer existed.

But perhaps the most compelling aspect of the Serpo document release was an extensive transcript of a briefing arranged on short notice for President Ronald Reagan in 1981.

Hidden Agenda

U.S. President Ronald Reagan giving the famous "alien" speech at the United Nations.

Of course, we've already discussed President Reagan's propensity for making remarks about aliens. This has led many to conclude that he had some inside knowledge of the U.S. government's involvement with extraterrestrials, including the Serpo information. What the transcript makes clear though is that there was a lot more to it than that. Reagan considered the alien question to be the most important he dealt with in his time in the White House.

Many claim that the Project Serpo documents are fake, but once again they've been out there for over a decade and no one has taken credit for the supposed ruse. What I find hard to dispute is the sheer volume of detail contained in the files, especially the Reagan transcript. Every player who is identified in the transcript speaks and behaves exactly as they were known to in real life. If it is a fake, the effort needed to create that air of authenticity is off the scale. Why would anyone bother, especially if they're giving away the information for free?

The transcript primarily records the interactions between President Reagan, CIA Director William Casey, and a mysterious figure known only as "The Caretaker," who is probably the then titular head of the MJ-12 organization. The briefing took place on March 6, 1981 at Camp David, and I think it is worth revisiting the text (italics are mine).

William Casey (CIA): "Mr. President, good morning.

The Project Serpo Human-Alien Exchange Program

As we discussed in February, this briefing contains some very sensational and some very, very classified information. This will be a real tough one to follow since the briefing starts back, historically speaking, that is, and runs up to recent times. I believe we have prepared a good chronological order of events. I'm sure you, Mr. President, will have many questions."

The Caretaker: "Good morning, Mr. President. First of all, I would like to give you a bit of information on my background. I have been employed by the CIA for the past 31 years. I started the caretaking status of this project in 1960. We have a special group of people whom we call "Group 6," that cares for all this information."

"The United States of America has been visited by Extraterrestrial Visitors since 1947. We have proof of that. However, we also have some proof that Earth has been visited for many *thousands of years* by various races of Extraterrestrial Visitors. Mr. President, I'll just refer to those visitors as ETs. In July, 1947, a remarkable event occurred in New Mexico. During a storm, two ET spacecraft crashed. One crashed southwest of Corona, New Mexico and one crashed near Datil, New Mexico. The U.S. Army eventually found both sites and recovered all of the debris and one live Alien. I'll refer to this live Alien as 'EBE 1.'"

President Reagan: "Can we classify them? I mean can we... well, connect them with anything Earthly?"

The Caretaker: "No, Mr. President. They don't have any similar characteristics of a human, with the exception of their having eyes, ears and a mouth. Their internal body organs are different. Their skin is different, their eyes, ears and even breathing is different. Their blood wasn't red and their brain was entirely different from human. We could not classify any part of the Aliens with humans. They had blood and skin, although considerably different than human skin. Their eyes had two different eyelids, Probably because their home planet was very bright.

"EBE stayed alive until 1952 when it died. We learned a great deal from EBE. EBE was extremely intelligent. It learned English quickly, mainly by listening to the military personnel who were responsible for EBE's safety and care. EBE showed us how some of the items we recovered worked, such as a communications device. It also showed us how various other devices worked.

"EBE died of what military doctors considered natural causes. I don't think we could really state exactly why EBE died. EBE did explain where he lives in the universe. We call this star system Zeta Reticuli, which is about 40 light-years from Earth. EBE's planet was within this star system."

Reagan: "How did they make such an incredibly long journey?"

The Caretaker: "It took the EBE spaceship nine of our months to travel the 40 light-years. Now, as you can see, that would mean the EBE spaceship traveled faster than the speed of light. But, this is where it gets really technical. Their spaceships can travel through a form of 'space tunnels' that gets them from point 'A' to point 'B' faster without having to travel at the speed of light."

Reagan: "OK, understood."

The Caretaker: "The original project, started back in 1947, was called 'Project GLEEM.' This project contained volumes of documented information collected from the beginning of our investigation of UFOs and Identified Alien Craft. The project was originally established in the early '50s by, first President Truman and then by order of President Eisenhower, under control of the National Security Council. President Truman established a group of people to handle this project. The group was called Majority 12 or 'MJ-12.'"

"In 1966, the project's name was changed to 'Aquarius.' The project was funded by confidential funds appropriated within the intelligence community's budget. The recovery of these alien spacecraft led the United States on an extensive investigative program to determine whether these aliens posed a direct threat to our National Security. The United States felt relatively sure the aliens' exploration of Earth was non-aggressive and non-hostile. It was also established that the aliens' presence did not directly threaten the security of the United States.

Reagan: "How was that determined?"

The Caretaker: "We were involved in one major operation during this timeframe that involved our alien visitors."

Reagan: "You mean, we were in direct contact with them?"

Casey: "Yes Mr. President..."

As we have seen, President Reagan was given a National Security briefing on the U.S. government's history with aliens and Project Serpo. What he had not been told yet was just how deep the rabbit hole went...

The Caretaker: "Many reported sightings and incidents have occurred over nuclear weapons bases. The Air Force has initiated measures to assure the security of the nuclear weapons from alien theft or destruction. MJ-12 feels confident that the aliens are on an exploration of our solar system for peaceful purposes. However, we do have information and that is at another level, that more than one alien species are visiting Earth."

Reagan: "You mean, an alien species that isn't on a peaceful exploration?"

Casey: "Possibly, Mr. President."

Reagan: "Well, I have a lot of questions, so let me ask a few and then we can move on. I guess the first question I have is their life span. How old is EBE-1?"

The Caretaker: "Mr. President, the alien civilization that EBE came from is what we call the Eben Society. It wasn't a name they gave us; it was a name we chose. Their life span is between 350-400 years, but that is Earth years."

Reagan: "Is time the same on their planet as on ours?"

The Caretaker: "No, Mr. President, time is very different on the Eben Planet, which, by the way, we call SERPO. Their day is approximately 40 hours. That is measured by the movement of their two suns. The solar system containing SERPO is a binary star system, or two suns, rather than one, like our solar system."

Reagan: "I see. Please continue."

Casey: "Mr. President, the Soviet Union has also had their contacts with these aliens. We have a great deal of intelligence that would indicate the Soviets had their 'Roswell,' so to speak. What they know is about the same as we know. They had some bodies back in the late '50s, but our intelligence would indicate the species of aliens were different."

Reagan: "OK, well, then Bill, that presents a very disturbing feeling for me. Are you telling me there are different races or species, as you said, visiting Earth at the same time?"

The Caretaker: "Yes, Mr. President, but I hesitate to state the reason. We should have that discussion in a different meeting."

Reagan: "Just answer the simple question. How many different species have visited us?"

The Caretaker: "Mr. Director?"

Casey: "Go ahead, Caretaker, answer the President's question."

The Caretaker: "At least five."

Reagan: "Are they all friendly?"

The Caretaker: "We have little intelligence on four of the five. We have plenty of intel on the Ebens. They have also helped us to understand the other four species. I'm afraid to say, Mr. President, and please don't misunderstand my words, but we think *one of the species is very hostile.*"

Casey: "Mr. President, do you wish for us to continue on this track or would you like something more private?"

Reagan: "For Christ sakes, I'm the President of the United States! I should know if we are endangered by some *threat from outer space.* If you have something to say about a threat posed by this one species of aliens, then *I want to hear it.*"

The Project Serpo Human-Alien Exchange Program

At this point, the briefing seems to confirm many of the points previously discussed in this book, especially the more ominous ones. And Reagan is about to become very alarmed at what he's hearing.

> Casey: "Mr. President, we have intelligence that would indicate this one species of aliens have *abducted people from Earth*. They have performed scientific and medical tests on these humans. To the best of our knowledge, no humans have been killed. We have captured one of these hostile aliens."
>
> Reagan: "Do we have *operational war plans* on this?"
>
> Casey: "Yes, Mr. President, we have war plans on *all* potential threats to our country."

The next section of the briefing sounds like something right out of the X-Files...

> The Caretaker: "There is no way we can control visitors from outer space from traveling to Earth and visiting our planet. We know they have visited us in the past and will visit us in the future. We must understand that the visitors can roam our planet at will without us doing much about it. However, I personally believe that we must prepare for the *eventual day when some hostile life form decides to take over our planet*."
>
> Reagan: "My God..."
>
> Casey: "Mr. President, the five species are called, Ebens, Archquloids, Quadloids, Heplaloids and Trantaloids. These names were given to the alien's species by the intelligence community, specifically MJ-5. The Ebens are friendly; the Trantaloids are the dangerous ones."
>
> Reagan: "Just knowing we have names for these things is amazing. Which one did we capture?"
>
> Casey: "Mr. President, we have a Trantaloid, but it is dead. We captured it in 1961 in Canada and we had it in captivity until 1962, when it died."

Hidden Agenda

The briefing then turns to the existence of a program that fits the designs of Brookings Report, to keep the public confused as to the reality of aliens and UFOs.

> The Caretaker: "In order to protect all this information and the fact that the United States Government has evidence of our planet being visited by Extraterrestrials, we developed over the years a very effective program to safeguard the information. We call it 'Project DOVE.' It is a complex series of disinformation operations by our military intelligence agencies to disinform the public.
>
> "The *first person* who helped us with this disinformation program, *Mr. George Adamski*, back in the early '50s, and up to all of the movie productions of UFO-related movies."
>
> Reagan: "I always knew there was some form of cooperation between our government and the motion picture industry. I heard rumors over the years... even during my acting days."
>
> The Caretaker: "Well, Mr. President, the first cooperative venture was the movie, *The Day the Earth Stood Still*. That was a cooperative venture with the United States Air Force and the movie industry."
>
> Reagan: "That movie, *Close Encounters*, was that one of them?"
>
> The Caretaker: "Yes, Mr. President, we provided the basic subject matter for that movie."

This entire transcript (and there is much more of it) is fascinating because it confirms most of what we already knew about the secret space program, including the project of cultural conditioning first mentioned in the Brookings Report. Reagan is known to have made mention of several incidents where the government leaked stories to Hollywood to put certain ideas into the heads of the American public.

There's a pretty well-vetted story to the effect that after a screening of *Close Encounters of the Third Kind* at the Reagan White House that President Reagan came up to Steven Spielberg and told him, "You don't know how true this really is."

The Project Serpo Human-Alien Exchange Program

A lot of the details of *Close Encounters* align with the Project Serpo releases. For instance, in the film the exchange team is made up of 12 people, 10 men and 2 women, just like the Project Serpo documents. The question is, where did Spielberg get that specific count? The Caretaker told Reagan flat out that if came from MJ-12.

We've had similar kinds of leaks since then as well. There was a really astonishing incident in 2003 when John Glenn went

John Glenn, the first American to orbit the Earth.

Hidden Agenda

on national TV and all but spilled the beans on being forced to lie about what he and the other astronauts had seen. The context was the most popular show on TV at the time, *Frasier* starring Kelsey Grammer as psychologist with a call-in radio show. Glenn had asked to make an appearance on the TV show. With Glenn playing himself, at one point Frasier steps away from the radio mic and Glenn begins to speak into it, directly to the American people.

> Back in those glory days, I was very uncomfortable when they asked us to say things we didn't want to say and deny other things. Some people asked, you know, were you alone out there? We never gave the real answer, and yet we see things out there, strange things, but we know what we saw out there. And we couldn't really say anything. The bosses were really afraid of this, they were afraid of the "War of the Worlds" type stuff, and about panic in the streets. So we had to keep quiet. And now we only see these things in our nightmares or maybe in the movies, and some of them are pretty close to being the truth.

One of the things that makes this statement so compelling to me is that it was obviously not scripted. The dialog is awkward and personal, and no writer would write such clumsy dialog even for a scene like this. It's clear to me that Glenn is ad-libbing the entire amazing statement in his own words.

So in one fell swoop, on the most popular TV show in the country, national hero John Glenn confirms not only that the Brookings Report recommendations were official NASA policy, but also that *he* has lied, that *NASA* has lied, that he and his colleagues were so shattered by what they saw that they have had "nightmares" ever since, and that some film portrayals about the UFO subject are accurate. That's amazing!

But the Serpo documents confirmed much more than just the existence of "Project DOVE." They also shed light on a frightening and chilling incident that took place at Area 51 in 1983. According to the Project Serpo sources, an Eben mechanic named J-Rod was assigned to help the United States figure out how to operate Eben technology. Along with J-Rod, an Archquloid (a genetic

The Project Serpo Human-Alien Exchange Program

automaton created by the Ebens), was also provided for American study. After more than a decade in captivity, J-Rod began to feel sorry for the Archquloid and helped it escape.

Apparently, the Archquloid wasn't treated very well by our scientists, and after years and years of captivity and experiments, it telepathically communicated to J-Rod that it wished to be free. J-Rod felt sorry for the creature, and led it to an armory where it was a given an alien weapon that was something like a particle beam or railgun. The alien then basically just walked out of the facility, and wasn't challenged until it reached the Gate 3 security checkpoint at Area 51.

As the story goes, a team of guards drove out to Gate 3 for a normal check-in with the guard, and they found the guard gone. There was blood inside the guard shack, and various pieces of a human body scattered about. The gate was broken and the guards began a search in their jeep. The guards found the Archquloid about a mile from the gate, and they confronted him, ordering him to surrender. The alien turned and pointed something at them, probably the railgun. When he did, the captain opened fire and hit the alien four times in the chest. The alien was taken to a medical facility and recovered, but died about a year later from complications associated with the confrontation. J-Rod was never trusted again and died in captivity many years later.

The violence associated with these confrontations, as well as the story of the Dulce, New Mexico battle, indicates that the relationships between the humans and aliens has been tense, at best. Yet, none of these leaks really proves anything without some sort of supporting evidence. That did not come until 2002.

In 2002 a Scottish systems administrator named Gary McKinnon began hacking into the U.S. Department of Defense mainframes. While poking around in there, he discovered numerous files and databases linked to the existence of a secret space program. What he found were photos of UFOs going back decades, along with documents pertaining to a large space fleet and personnel. These personnel files contained lists of what were called "non-terrestrial" officers. They weren't assigned to any branch of the U.S. military, at least not any official branch. The documents also contained manifests for supply transfers from ship to ship in this secret space fleet.

McKinnon was eventually caught and charged with damaging files and equipment owned by the U.S. government. They attempted to extradite him for prosecution in the U.S., but in 2012 the British government denied the request and as of the moment, McKinnon is free.

McKinnon's discoveries all but confirmed the existence of the secret space program, but the attempts to prosecute him are seen by some as an attack against the so-called "disclosure movement."

There are some who claim that the U.S. government is trying to get out of the UFO business, and wants to disclose the alien activity to the public at large.

Honestly I'm pretty skeptical of the so-called "disclosure movement." For the better part of 15 years, I've been hearing that next week, next month or after the next election the government is going to come clean about what they know about UFOs. It never happens.

A few years ago in Washington, D.C. Steve Basset and the disclosure crowd held a "people's congressional hearing" on disclosure to try to apply pressure on the government. The event was completely ignored by the media and didn't generate any additional leaks about aliens or the secret space program.

To me, the whole notion of official disclosure is very naive. If you are the president of the United States, you're already at the top of the political food chain. If you introduce a destabilizing element into the body politic—which alien disclosure certainly is—the results are completely unpredictable. Pretty much the only direction your career is going to go is down. Your political life is over. I can't see anyone in a position of authority taking that risk.

I also don't see why anyone cares or wants that kind of sanction. I mean, we already know that the government has lied to us for over 60 years on this subject. Why do we assume that what they "disclose" is anything remotely close to the truth? And why do we need the approval of these lying bastards at all? Does the disclosure crowd really think that if the president reveals the truth about ET that they'll all be on TV talk shows the next day? I just don't see that happening.

The Project Serpo Human-Alien Exchange Program

But there is one circumstance under which such a revelation might be beneficial to the leaders that have lied to the world for more than a half century.

Personally, if there ever is disclosure in some speech from the Oval Office, with the pictures of the kids and the paned windows in the background, I'm heading for the hills. Because that means that our leaders have nothing to lose politically anymore. It means that something very, very bad is coming. And I don't think any of us wants to live through that.

I think that the only disclosure we're going to get is the same kind we've seen over the years. Leaks here and there, the odd whistleblower, and the occasional hint in our movies or TV shows.

In a way, we should be comforted by the secret space program. It means that somebody has analyzed the threat and tried to build some defenses against it. It also shows that the Brookings policies are still in effect. That they are still worried about worldwide panic if these facts come out.

But what about the other part of Brookings? If the secret space program exists, and if it possesses technologies far beyond what we see today, isn't it possible we have visited Mars or the Moon, and perhaps even have bases there?

Yes, it certainly is.

Hidden Agenda

Wernher von Braun poses next to the engines of a Saturn rocket.

Chapter 9
Andromedans, Blue Avians and the Hollow Earth

I cannot end this book without addressing some of the wilder claims I've seen made in the whole "secret space program" industry lately. As we all know, for a fee, we can join a web site here and there and pay for premium content presented by well-known "experts" on the secret space program and watch literally hours of content about all aspects of it. Invariably, these wild tales are "leaks" of fantastic information from "insiders" who have seen the workings of the secret space program up close. They've been to secret U.S. government bases inside mountains at Area 51, or deep underground beneath Sedona, Arizona or in dark reptilian strongholds deep in the Black Hills of South Dakota where missing children are sexually abused and then eaten by evil aliens.

Others tell stories of being selected for special duty at an early age and recruited to travel the galaxy to speak for Earth at the Galactic Federation Council. They tell us about the kind Nordic races and the gentle "Blue Avians" from Andromeda who are only here to observe and assist our ascension to a higher form of consciousness. In a few hundred years, they promise, we will all become galactic beings of light and join the higher brotherhood of the galactic community where we will grovel and apologize for the error of our ways in the destruction of Earth's environment and our judgment and greed in taking advantage of our fellow man.

Yeah right.

If this sounds like the usual snake oil salesman's pitch, that's because it is. For every member of a "dark cabal" that's trying to steal our souls and sell our children to the Reptilians, trust me, there are equally powerful forces out there fighting them. These silly stories are presented with no evidence whatsoever, yet still the UFO community laps it up like turkey gravy at Thanksgiving.

Hidden Agenda

    I contemplated whether I should take these charlatans on directly in this book or not. The easiest thing, in terms of my career, such as it is, would be to just whistle on by and let them keep making their absurd claims and hope that the UFO community wises up and stops buying what they are selling. But I am a man driven by what I know to be the truth, and it is only by speaking that truth that I can sleep at night.

    So here we go.

    For starters, the Earth is not in peril. It never has been. The Earth has chugged along quite nicely for the last 4-billion years or so and shows no signs of collapsing anytime soon. Mankind's influence on the planet is minimal, at best. The oceans are not dying, the planet is not warming, and processed foods and pesticides are not killing us. If you don't believe me, go to the beach sometime and stand at the edge and look out to the horizon. Compare yourself to the tiny fraction of ocean you can observe from your vantage point. It's a great way to appreciate just how small each of us is. If I dedicated my entire life's energy to destroying the environment of the planet I couldn't do it— not in 10 thousand lifetimes. And neither can you.

    What science has taught us recently is that the sun (imagine that!) dictates whether the planet warms or cools, not a harmless (if not beneficial) trace gas named carbon dioxide that makes up less than .004 of the total atmospheric volume of the planet. It has taught us that the human condition continues to improve at a staggering pace. Each year, our crops yield more while we feed more people, infant mortality rates plummet and lifespans expand. In short, the snake oil salesmen who want to tell you that the Earth is going to hell in a handbasket next week are just trying to scare you so you'll buy into their vision of the future.

    And I do mean "buy."

    The first of these charlatans to emerge was Andrew D. Basiago, an American lawyer from Vancouver, Washington who also refers to himself as a "chrononaut, and 21st century visionary." He claims to have been born on September 18, 1961 in Morristown, New Jersey, and grew up in northern New Jersey and southern California. He says he was "identified in early childhood as an 'Indigo' child with special abilities, including the ability to

Andromedans, Blue Avians & the Hollow Earth

use his mind to levitate small objects and to perform telepathy by reading the minds of others." He also claims to be a past member of Mensa and that he holds five degrees, including a BA in History from the University of California at Los Angeles (UCLA) and a Master of Philosophy from the University of Cambridge.

Basiago became a fixture in the UFO community when he began talking about his alleged involvement with something called Project Pegasus. According to him, Project Pegasus was a DARPA (Defense Advanced Research Projects Agency) program to develop teleportation technology. Basiago says that he was teleported to Mars from a TRW facility in Redondo Beach, California numerous times as child in from 1967 forward. He claims that he was teleported to Mars where he saw an extensive American base of operations there. He also says that he travelled both backward and forward in time, even attending Lincoln's Gettysburg address. Oh, and one of his time and space traveling buddies during this period was Barack Obama, then operating under his pseudonym "Barry Sotero."

Of course, he can provide no evidence for any of this.

I worked for several years during my previous incarnation as an aerospace engineer at Space Park in Redondo Beach, California. In fact, I worked at the old TRW building just a few hundred feet from the building where Basiago claims he teleported to Mars. I can assure you, having held several security clearances, that there is no teleporter in that building. The bathrooms even have substandard plumbing.

It is my opinion, after reading Basiago's claims, that he's simply a disturbed man who should be gently nudged toward seeking psychological help. But what disturbs me even more is that so many in the UFO community actually took him seriously for such a long time.

This brings us to the most recent "whistleblower" on the secret space program, Corey Goode.

Goode calls himself "an intuitive empath (IE) with precognitive abilities," and claims he was identified and abducted at an early age through something he calls the MILAB (military abduction training and indoctrination) program. He claims he "served and trained" in the MILAB program beginning at the age

181

of six from 1976 to 1987. Toward the end of his time as a MILAB he was assigned to an IE support role for a rotating Earth Delegate Seat (shared by secret earth government groups) in a "human-type" ET "Super Federation Council."

Goode's website goes on to claim that his IE abilities:

> ...played an important role in communicating with non-terrestrial beings" (termed 'interfacing') as part of one of the Secret Space Programs (SSP). During his 20 year service he had a variety of experiences and assignments including the Intruder Intercept Interrogation Program, Assignment to the ASSR-ISRV (Auxiliary Specialized Space Research, Interstellar Class Vessel), and much more. This all occurred in a '20 and Back' agreement from 1986/87-2007 with recall work until the present day.

What he means by "20 and back" is that he served humanity for 20 years on the ET Super Federation Council from 1987 to 2007, and then was "reverse aged" to be given his 20 years back.

Yeah. Sure.

His website then goes on to talk about what Goode is up to since his "retirement" from the Super Federation Council in 2007.

> Goode now works in the information technology and communications industry with 20 years' experience in hardware and software virtualization, physical and IT security, counter electronic surveillance, risk assessment, and executive protection, and served in the Texas Army State Guard (2007-2012), C4I (Command, Control, Communications, Computation & Intelligence). The time in the Texas Military Forces was unrelated to the Secret Space Program Service.

Ok, correct me if I'm wrong here, but if he has 20 years' experience in the IT world, wouldn't that make an overlap of 10 years where he was serving the Super Federation Council while simultaneously working his day job as the tech support guy? Would that be kind of hard to do, assuming that Super Federation

Andromedans, Blue Avians & the Hollow Earth

Corey Goode.

Council held its meetings off world? Unless they rented out the local VFW hall to hold their meetings...

But it doesn't stop there. According to his site:

> Goode continues his IE work now and is in direct physical contact with the Blue Avians (of the Sphere Being Alliance) who have chosen him as a delegate to interface with multiple ET Federations and Councils on their behalf, liaison with the SSP Alliance Council, and to deliver important messages to humanity.

Uh right. So this guy is ex-Super Federation Council, and currently serving as a delegate to the Sphere Being Alliance, and "multiple ET Federations and Councils" and the best gig the Blue Avians can find him is an IT job? I mean really, what happens if the Sphere Being Alliance holds a meeting on Alpha Centuri the same day he has to finish that Windows 10 malware security update at the Encino corporate offices? Is there an Uber service to Alpha Centuri? And who pays the fee?

Overlooking the apparent inconsistencies in his stories, one might still ask, what proof does he offer of any of his claims?

Absolutely nothing.

Hidden Agenda

Goode is also part of the "climate change" crowd, using fear of catastrophic weather changes and global warming as a wedge to convince people that our space brothers will save us. Once again this is totally at odds with the facts. The Earth has seen no real increase in surface temperatures in the last 20 years or so, despite the constant slew of "hottest year on record" claims made by NASA and NOAA. The satellite data prove that. In fact, it appears we are heading for a substantial cooling period over the next 25 to 50 years.

There is also the fact that the climate of the Earth hasn't changed at all in our lifetimes. In 1960, southern California was sunny and dry. Today, it is sunny and dry. In 1960, the Pacific Northwest was wet and mild. Today, it is wet and mild. In 1960, Florida was hot and humid. Today, it is hot and humid. In 1960, the upper Midwest was hot and humid in the summer, and cold and snowy in the winter. Today it is hot and humid in the summer, and cold and snowy in the winter. In 1960, the Great Plains were windy and stormy. Today, they are windy and stormy. The only measureable changes in the general climate since the 1950s is that there are fewer severe storms, hurricanes and tornados, which is an indication of cooler temperatures rather than warmer temperatures.

So Goode's alien friends seem to address a condition which doesn't exist.

There is also the matter of the Hollow Earth. According to Goode, the Earth is inhabited in its interior by multiple races of beings, some of whom developed in the "honeycomb Earth" and some of whom migrated there. He also says that there is a great deal of highly coveted "multi-dimensional" technology left behind by a "builder race" that assembled the Moon and other bodies in the solar system. Again, he offers no proof. Frankly, if he is really in contact with extraterrestrial races I assume they could show him some of the entrances to the "honeycomb Earth" and perhaps he could conduct tours of some of the upper levels for a few extra bucks. After all, a guy's gotta eat…

Now that I've slammed him, I must say that I have no problem with Goode or anyone else making a living from writing books, talking about the Hollow Earth (I've heard the same stories) or hitting the lecture circuit. I simply want to point out that there must be some standards for evaluating such claims, and whatever they are Goode doesn't meet them. At least Robert O. Dean, Clifford Stone and Bob Lazar have some documentary evidence of what they claim. Basiago and Goode have none.

To his credit, Goode says that he is doing this all for the *Goode* of humanity, and that his Blue Avian friends have a message for mankind that he is destined to convey. So, what is the message? According to his website:

> Every day focus on becoming more "Service To Others" oriented. Focus on being more "Loving" and "Focus on raising your Vibrational and Consciousness Level" and to learn to "Forgive Yourself and Others" (Thus "Releasing Karma"). This will change the Vibration of the Planet, The "Shared Consciousness of Humanity" and "Change Humanity One Person at a time" (Even if that "One Person" is yourself). They say to treat your body as a temple and change over to a "Higher Vibrational Diet" to aid in the other changes. This sounds to many like a "Hippy Love and Peace" message that will not make a difference. I assure you the "Path" they lay out in "Their Message" is a difficult one. Even on the unlikely chance that these technologies stay "Suppressed", imagine what a world we would live in if everyone made these changes to their selves?

This all sounds great—as he puts it, like some happy hippie crap that we've all heard before. But to me, "service to others" rings like a call to suppress your individual needs to some greater whole. This is a seductive groupthink message popular with the globalist elite. It is akin to telling you that the ridiculous tax rates you pay to "pay for" a government that is privately funded and can simply print any amount of money they need is a "patriotic duty." The reality is that it is our uniqueness as individuals that separates us from the other lockstep clone races like the Greys, and honestly I think that's the last thing we should discard. It's what makes us special.

To my mind, Goode's message is a clarion call to abandon

that which makes us uniquely human, and far from being a step forward for humanity it instead strips us of our power and exceptionalism. I say stay away.

A publisher of mine once told me that catastrophe sells, and to my mind that is what these purveyors of "dark cabals" and "death by ascension" advocate. I urge you not to buy into it (literally) but instead focus on what is real and has some evidence to support it. The truth of the existence and purpose of the secret space program may not be as sexy and spiritually satisfying as teleportation, time travel and Blue Avian space brothers, but it is still a fascinating story that may give us hints at the true nature of the universe we live in.

***

In this volume, I have tried to concentrate on just those aspects of the history of the secret space program that I can support with at least some documentary evidence and logical deduction. In conclusion, I've created this timeline of events that I think best describes just what the secret space program is, how it works and where it came from. I don't expect you to necessarily agree with my conclusions, but I hope you will give them due consideration.

As near as I can tell, and as I have set forth in this volume, here is the not-so-secret history of the secret space program:

- 1800-1850—A secret society known as NYMZA or NJMZA forms in Prussia for the purpose of studying ancient texts through physicists' eyes. Intrigued by stories of Vimanas and other flying machines, these wealthy industrialists begin to recruit physicists and engineers to their cause.

- 1850—NYMZA sends an engineer named Charles Dellschau to the United States to conduct research and create prototypes of flying machines. He travels to Sonora, California and forms the Sonora Air Club, which has a breakthrough in 1860 with the discovery of the mysterious N/B Gas which has anti-gravity properties.

Andromedans, Blue Avians & the Hollow Earth

- 1896-97—A wave of airship sightings spreads across the U.S., starting near Sonora and following a path that seems to point back to Prussia.

- 1920s-1930s—Einstein discovers the concept of torsion, a twisting of spacetime which can "gate" energy from higher spatial dimensions through the rotation of matter. T. Townsend Brown publishes "How I Control Gravity" detailing his experiments with electrically charged disk shaped massive test articles. Viktor Schauberger tests his disk shaped "Repulsine" device for the first time. Einstein flees Nazi Germany leaving his research behind and abandons his study of torsion.

- 1940—Nazi Germany forms an advanced physics/weapons development team headed by SS General Hans Kammler. A huge research facility is constructed in Silesia, a region of East Prussia.

- 1942—A flying saucer crashes in Cape Girardeau, Missouri. From it, the U.S. government is able to eventually reverse engineer fiber optic technology.

- 1942—Spooked by war jitters and possibly the Cape Girardeau events, U.S. military forces engage in an hours-long "battle" with a flying disk over Los Angeles. The object is impervious to anti-aircraft fire.

- 1944—Nazi Germany begins experimentation on "Die Glocke"—the Nazi Bell. It uses massive counter-rotating magnets and a mysterious substance known as "Xerum 525" to amplify the effects. Xerum 525 is almost certainly a further development of the NYMZA N/B Gas or what is today known as "Red Mercury." The Bell is lethal to all biological life but produces a powerful ant-gravity effect.

- 1945—Nazi Germany falls and Hans Kammler and Die Glocke disappear. It is widely rumored

Hidden Agenda

that they both escaped to a secret Nazi base in Neuschwabenland, Antarctica to continue the Bell research.

- 1946—Operation Highjump, a virtual invasion Antarctica, is launched in the southern hemisphere summertime. The task forces surround and operate in Neuschwabenland. After barely a month, the mission is abandoned and tons of equipment is left behind. Rumors persist that the Highjump troops encountered a superior force of Nazi flying saucers and were routed. After returning, Admiral Richard Byrd warns in an interview that the U.S. must prepare to be attacked "from the poles" by craft capable of incredible speeds.

- June 1947—Kenneth Arnold spots a squadron of experimental flying wing aircraft based on the Nazi Horten HO 229 fighter. His sighting is misinterpreted and starts a wave of UFO sightings.

- July 1947—Something, probably an alien spacecraft, crashes near Roswell, New Mexico. Bodies and the craft are recovered and sent for reverse engineering to American industry. The transistor is the first major breakthrough from the find. The government forms a special committee known as Majestic or MJ-12. MJ-12 assumes all control over information pertaining to the alien presence on Earth.

- 1950-1978—Communication is established between the U.S. government and a race of "Grey" type beings called "Ebens" from Zeta Reticuli. A cultural and technological exchange program is established between the two worlds. Twelve Americans travel to the Eben homeworld, code named "Serpo." Eight of them return in 1978.

- July 1952—Washington, D.C. is "buzzed" over two consecutive weekends by craft displaying performance capabilities far beyond those of the American military. A panic ensues. Publically, Project Blue-Book is initiated. Privately, new government agencies like the NSA and DARPA are formed to address the alien threat.

- 1955-1958—The U.S. government and aerospace corporations become fascinated with anti-gravity technologies. T. Townsend Brown submits a proposal to the Navy for "Project Winterhaven," an anti-gravity flying disk craft based on his research. The Navy declines his proposal but then promptly takes all anti-gravity research into the black.

- 1960—President Kennedy assumes office. Finding himself locked out of the MJ-12 national security state, Kennedy decides to initiate his own secret space program under the cover of NASA. The true purpose of the Apollo program is to salvage alien technology left behind on the Moon and reverse engineer it to bring the "public" secret space program on a technological par with the MJ-12 group, which he has no control over.

- 1958-1963—Wernher von Braun discovers the influence of rotation on rocket performance and celestial navigation. The U.S. Army proposes Project Horizon: a secret base on the Moon to be built by 1966.

- November 1963—President John F. Kennedy reaches an agreement with the Soviet Union for a joint program to travel to the Moon. Ten days later he is assassinated.

- March 1981—In a special meeting at Camp David, President Reagan is briefed on the alien presence by CIA director William Casey and a mysterious figure known as "The Caretaker." Alarmed by information indicating that some alien races are hostile, Reagan

Hidden Agenda

demands war plan contingencies be generated. Twenty-four days later, he is shot in an assassination attempt by John Hinckley, a man with ties to the family of his vice president.

- 1983—Having returned to health and rebuilt the U.S. economy, Reagan launches the Strategic Defense Initiative, or "Star Wars" program, to develop space-based defense systems against alien attack. Reagan and future Soviet Premier Mikhail Gorbachev establish a secret agreement to share the technologies they separately develop at the Reykjavík summit in 1986.

- 1987—The existence of the Alien Reproduction Vehicle (ARV) is revealed at a military air show in California. It is based in part on the work of T. Townsend Brown, and partially on reverse engineered technology from crashed flying saucers of alien origin.

- 1990s—Two space shuttle missions, STS-48 and STS-80, record tests of the ARV in low Earth orbit.

- 2000s—"Jon Lenard Walson" and others take images of massive space platforms in orbit around the Earth that may be part of the planetary defense created under the Star Wars program.

- 2010s—Various "whistleblowers" telling fantastic tales of the secret space program pop up in various Internet and social media forums. None of their information can be corroborated. Gary McKinnon hacks into the DOD defense mainframe and discovers a list of "non-terrestrial officers" who appear to be astronauts within the secret space program.

That, sadly, is pretty much it. Perhaps in 20 years we will be able to look back, put the pieces together and realize that this was actually an extremely active period for the secret space program, but right now it's hard to see that happening. Information from credible sources seems to have dried up just when we have the means to communicate it better than ever before. I have no doubt that at least two separate and parallel secret space programs existed as recently as the 1960s, but since then things have gone very quiet. I suspect that around the time Reagan took power there was a major shift behind the scenes, and the two separate programs linked up for the good of the planet. Gary McKinnon's hack of the Department of Defense is really the only significant development in this decade.

Until we get more credible information, all we can do is watch, and wait.

And wonder...

Hidden Agenda

Fig. II-6. Typical Lunar Construction Vehicle

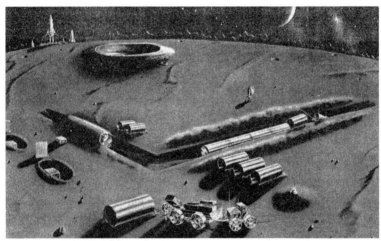

Above: Two illustrations from the U.S. Army's Project Horizon proposal.

Get these fascinating books from your nearest bookstore or directly from: Adventures Unlimited Press
www.adventuresunlimitedpress.com

DEATH ON MARS
The Discovery of a Planetary Nuclear Massacre
By John E. Brandenburg, Ph.D.

New proof of a nuclear catastrophe on Mars! In an epic story of discovery, strong evidence is presented for a dead civilization on Mars and the shocking reason for its demise: an ancient planetary-scale nuclear massacre leaving isotopic traces of vast explosions that endure to our present age. The story told by a wide range of Mars data is now clear. Mars was once Earth-like in climate, with an ocean and rivers, and for a long period became home to both plant and animal life, including a humanoid civilization. Then, for unfathomable reasons, a massive thermo-nuclear explosion ravaged the centers of the Martian civilization and destroyed the biosphere of the planet. But the story does not end there. This tragedy may explain Fermi's Paradox, the fact that the cosmos, seemingly so fertile and with so many planets suitable for life, is as silent as a graveyard.

278 Pages. 6x9 Paperback. Illustrated. Bibliography. Color Section. $19.95. Code: DOM

BEYOND EINSTEIN'S UNIFIED FIELD
Gravity and Electro-Magnetism Redefined
By John Brandenburg, Ph.D.

Brandenburg reveals the GEM Unification Theory that proves the mathematical and physical interrelation of the forces of gravity and electromagnetism! Brandenburg describes control of space-time geometry through electromagnetism, and states that faster-than-light travel will be possible in the future. Anti-gravity through electromagnetism is possible, which upholds the basic "flying saucer" design utilizing "The Tesla Vortex." Chapters include: Squaring the Circle, Einstein's Final Triumph; A Book of Numbers and Forms; Kepler, Newton and the Sun King; Magnus and Electra; Atoms of Light; Einstein's Glory, Relativity; The Aurora; Tesla's Vortex and the Cliffs of Zeno; The Hidden 5th Dimension; The GEM Unification Theory; Anti-Gravity and Human Flight; The New GEM Cosmos; more. Includes an 8-page color section.

312 Pages. 6x9 Paperback. Illustrated. $18.95. Code: BEUF

VIMANA:
Flying Machines of the Ancients
by David Hatcher Childress

According to early Sanskrit texts the ancients had several types of airships called vimanas. Like aircraft of today, vimanas were used to fly through the air from city to city; to conduct aerial surveys of uncharted lands; and as delivery vehicles for awesome weapons. David Hatcher Childress, popular *Lost Cities* author and star of the History Channel's long-running show Ancient Aliens, takes us on an astounding investigation into tales of ancient flying machines. In his new book, packed with photos and diagrams, he consults ancient texts and modern stories and presents astonishing evidence that aircraft, similar to the ones we use today, were used thousands of years ago in India, Sumeria, China and other countries. Includes a 24-page color section.

408 Pages. 6x9 Paperback. Illustrated. $22.95. Code: VMA

ANCIENT ALIENS & SECRET SOCIETIES
By Mike Bara
Did ancient "visitors"—of extraterrestrial origin—come to Earth long, long ago and fashion man in their own image? Were the science and secrets that they taught the ancients intended to be a guide for all humanity to the present era? Bara establishes the reality of the catastrophe that jolted the human race, and traces the history of secret societies from the priesthood of Amun in Egypt to the Templars in Jerusalem and the Scottish Rite Freemasons. Bara also reveals the true origins of NASA and exposes the bizarre triad of secret societies in control of that agency since its inception. Chapters include: Out of the Ashes; From the Sky Down; Ancient Aliens?; The Dawn of the Secret Societies; The Fractures of Time; Into the 20th Century; The Wink of an Eye; more.
288 Pages. 6x9 Paperback. Illustrated. $19.95. Code: AASS

THE CRYSTAL SKULLS
Astonishing Portals to Man's Past
by David Hatcher Childress and Stephen S. Mehler
Childress introduces the technology and lore of crystals, and then plunges into the turbulent times of the Mexican Revolution form the backdrop for the rollicking adventures of Ambrose Bierce, the renowned journalist who went missing in the jungles in 1913, and F.A. Mitchell-Hedges, the notorious adventurer who emerged from the jungles with the most famous of the crystal skulls. Mehler shares his extensive knowledge of and experience with crystal skulls. Having been involved in the field since the 1980s, he has personally examined many of the most influential skulls, and has worked with the leaders in crystal skull research, including the inimitable Nick Nocerino, who developed a meticulous methodology for the purpose of examining the skulls.
294 pages. 6x9 Paperback. Illustrated. Bibliography. $18.95. Code: CRSK

AXIS OF THE WORLD
The Search for the Oldest American Civilization
by Igor Witkowski
Polish author Witkowski's research reveals remnants of a high civilization that was able to exert its influence on almost the entire planet, and did so with full consciousness. Sites around South America show that this was not just one of the places influenced by this culture, but a place where they built their crowning achievements. Easter Island, in the southeastern Pacific, constitutes one of them. The Rongo-Rongo language that developed there points westward to the Indus Valley. Taken together, the facts presented by Witkowski provide a fresh, new proof that an antediluvian, great civilization flourished several millennia ago.
220 pages. 6x9 Paperback. Illustrated. References. $18.95. Code: AXOW

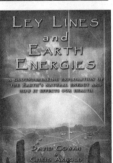

LEY LINE & EARTH ENERGIES
An Extraordinary Journey into the Earth's Natural Energy System
by David Cowan & Chris Arnold
The mysterious standing stones, burial grounds and stone circles that lace Europe, the British Isles and other areas have intrigued scientists, writers, artists and travellers through the centuries. How do ley lines work? How did our ancestors use Earth energy to map their sacred sites and burial grounds? How do ghosts and poltergeists interact with Earth energy? How can Earth spirals and black spots affect our health? This exploration shows how natural forces affect our behavior, how they can be used to enhance our health and well being.
368 PAGES. 6x9 PAPERBACK. ILLUSTRATED. $18.95. CODE: LLEE

ANCIENT ALIENS ON THE MOON
By Mike Bara
What did NASA find in their explorations of the solar system that they may have kept from the general public? How ancient really are these ruins on the Moon? Using official NASA and Russian photos of the Moon, Bara looks at vast cityscapes and domes in the Sinus Medii region as well as glass domes in the Crisium region. Bara also takes a detailed look at the mission of Apollo 17 and the case that this was a salvage mission, primarily concerned with investigating an opening into a massive hexagonal ruin near the landing site. Chapters include: The History of Lunar Anomalies; The Early 20th Century; Sinus Medii; To the Moon Alice!; Mare Crisium; Yes, Virginia, We Really Went to the Moon; Apollo 17; more. Tons of photos of the Moon examined for possible structures and other anomalies.
248 Pages. 6x9 Paperback. Illustrated.. $19.95. Code: AAOM

ANCIENT ALIENS ON MARS
By Mike Bara
Bara brings us this lavishly illustrated volume on alien structures on Mars. Was there once a vast, technologically advanced civilization on Mars, and did it leave evidence of its existence behind for humans to find eons later? Did these advanced extraterrestrial visitors vanish in a solar system wide cataclysm of their own making, only to make their way to Earth and start anew? Was Mars once as lush and green as the Earth, and teeming with life? Chapters include: War of the Worlds; The Mars Tidal Model; The Death of Mars; Cydonia and the Face on Mars; The Monuments of Mars; The Search for Life on Mars; The True Colors of Mars and The Pathfinder Sphinx; more. Color section.
252 Pages. 6x9 Paperback. Illustrated. $19.95. Code: AMAR

ANCIENT ALIENS ON MARS II
By Mike Bara
Using data acquired from sophisticated new scientific instruments like the Mars Odyssey THEMIS infrared imager, Bara shows that the region of Cydonia overlays a vast underground city full of enormous structures and devices that may still be operating. He peels back the layers of mystery to show images of tunnel systems, temples and ruins, and exposes the sophisticated NASA conspiracy designed to hide them. Bara also tackles the enigma of Mars' hollowed out moon Phobos, and exposes evidence that it is artificial. Long-held myths about Mars, including claims that it is protected by a sophisticated UFO defense system, are examined. Data from the Mars rovers Spirit, Opportunity and Curiosity are examined; everything from fossilized plants to mechanical debris is exposed in images taken directly from NASA's own archives.
294 Pages. 6x9 Paperback. Illustrated. $19.95. Code: AAM2

ANCIENT TECHNOLOGY IN PERU & BOLIVIA
By David Hatcher Childress
Childress speculates on the existence of a sunken city in Lake Titicaca and reveals new evidence that the Sumerians may have arrived in South America 4,000 years ago. He demonstrates that the use of "keystone cuts" with metal clamps poured into them to secure megalithic construction was an advanced technology used all over the world, from the Andes to Egypt, Greece and Southeast Asia. He maintains that only power tools could have made the intricate articulation and drill holes found in extremely hard granite and basalt blocks in Bolivia and Peru, and that the megalith builders had to have had advanced methods for moving and stacking gigantic blocks of stone, some weighing over 100 tons.
340 Pages. 6x9 Paperback. Illustrated.. $19.95 Code: ATP

COVERT WARS & THE CLASH OF CIVILIZATIONS
UFOs, Oligarchs and Space Secrecy
By Joseph P. Farrell
Farrell's customary meticulous research and sharp analysis blow the lid off of a worldwide web of nefarious financial and technological control that very few people even suspect exists. He elaborates on the advanced technology that they took with them at the "end" of World War II and shows how the breakaway civilizations have created a huge system of hidden finance with the involvement of various banks and financial institutions around the world. He investigates the current space secrecy that involves UFOs, suppressed technologies and the hidden oligarchs who control planet earth for their own gain and profit.
358 Pages. 6x9 Paperback. Illustrated. $19.95. Code: CWCC

ROSWELL AND THE REICH
The Nazi Connection
By Joseph P. Farrell
Farrell has meticulously reviewed the best-known Roswell research from UFO-ET advocates and skeptics alike, as well as some little-known source material, and comes to a radically different scenario of what happened in Roswell, New Mexico in July 1947, and why the US military has continued to cover it up to this day. Farrell presents a fascinating case sure to disturb both ET believers and disbelievers, namely, that what crashed may have been representative of an independent postwar Nazi power—an extraterritorial Reich monitoring its old enemy, America, and the continuing development of the very technologies confiscated from Germany at the end of the War.
540 pages. 6x9 Paperback. Illustrated. $19.95. Code: RWR

SECRETS OF THE UNIFIED FIELD
The Philadelphia Experiment, the Nazi Bell, and the Discarded Theory
by Joseph P. Farrell
Farrell examines the now discarded Unified Field Theory. American and German wartime scientists and engineers determined that, while the theory was incomplete, it could nevertheless be engineered. Chapters include: The Meanings of "Torsion"; Wringing an Aluminum Can; The Mistake in Unified Field Theories and Their Discarding by Contemporary Physics; Three Routes to the Doomsday Weapon: Quantum Potential, Torsion, and Vortices; Tesla's Meeting with FDR; Electromagnetic Phase Conjugations, Phase Conjugate Mirrors, and Templates; The Unified Field Theory, the Torsion Tensor, and Igor Witkowski's Idea of the Plasma Focus; tons more.
340 pages. 6x9 Paperback. Illustrated. $18.95. Code: SOUF

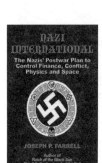
NAZI INTERNATIONAL
The Nazi's Postwar Plan to Control Finance, Conflict, Physics and Space
by Joseph P. Farrell
Beginning with prewar corporate partnerships in the USA, including some with the Bush family, he moves on to the surrender of Nazi Germany, and evacuation plans of the Germans. He then covers the vast, and still-little-known recreation of Nazi Germany in South America with help of Juan Peron, I.G. Farben and Martin Bormann. He then covers Nazi Germany's penetration of the Muslim world including Wilhelm Voss and Otto Skorzeny in Gamel Abdul Nasser's Egypt before moving on to the development and control of new energy technologies including the Bariloche Fusion Project, Dr. Philo Farnsworth's Plasmator, and the work of Dr. Nikolai Kozyrev. Finally, he discusses the Nazi desire to control space, and examines their connection with NASA, the esoteric meaning of NASA Mission Patches.
412 pages. 6x9 Paperback. Illustrated. $19.95. Code: NZIN

TECHNOLOGY OF THE GODS
The Incredible Sciences of the Ancients
by David Hatcher Childress

Childress looks at the technology that was allegedly used in Atlantis and the theory that the Great Pyramid of Egypt was originally a gigantic power station. He examines tales of ancient flight and the technology that it involved; how the ancients used electricity; megalithic building techniques; the use of crystal lenses and the fire from the gods; evidence of various high tech weapons in the past, including atomic weapons; ancient metallurgy and heavy machinery; the role of modern inventors such as Nikola Tesla in bringing ancient technology back into modern use; impossible artifacts; and more.
356 PAGES. 6x9 PAPERBACK. ILLUSTRATED. $16.95. CODE: TGOD

COVERT WARS AND BREAKAWAY CIVILIZATIONS
By Joseph P. Farrell

Farrell delves into the creation of breakaway civilizations by the Nazis in South America and other parts of the world. He discusses the advanced technology that they took with them at the end of the war and the psychological war that they waged for decades on America and NATO. He investigates the secret space programs currently sponsored by the breakaway civilizations and the current militaries in control of planet Earth. Plenty of astounding accounts, documents and speculation on the incredible alternative history of hidden conflicts and secret space programs that began when World War II officially "ended."
292 Pages. 6x9 Paperback. Illustrated. $19.95. Code: BCCW

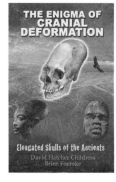

THE ENIGMA OF CRANIAL DEFORMATION
Elongated Skulls of the Ancients
By David Hatcher Childress and Brien Foerster

In a book filled with over a hundred astonishing photos and a color photo section, Childress and Foerster take us to Peru, Bolivia, Egypt, Malta, China, Mexico and other places in search of strange elongated skulls and other cranial deformation. The puzzle of why diverse ancient people—even on remote Pacific Islands—would use head-binding to create elongated heads is mystifying. Where did they even get this idea? Did some people naturally look this way—with long narrow heads? Were they some alien race? Were they an elite race that roamed the entire planet? Why do anthropologists rarely talk about cranial deformation and know so little about it? Color Section.
250 Pages. 6x9 Paperback. Illustrated. $19.95. Code: ECD

ARK OF GOD
The Incredible Power of the Ark of the Covenant
By David Hatcher Childress

Childress takes us on an incredible journey in search of the truth about (and science behind) the fantastic biblical artifact known as the Ark of the Covenant. This object made by Moses at Mount Sinai—part wooden-metal box and part golden statue—had the power to create "lightning" to kill people, and also to fly and lead people through the wilderness. The Ark of the Covenant suddenly disappears from the Bible record and what happened to it is not mentioned. Was it hidden in the underground passages of King Solomon's temple and later discovered by the Knights Templar? Was it taken through Egypt to Ethiopia as many Coptic Christians believe? Childress looks into hidden history, astonishing ancient technology, and a 3,000-year-old mystery that continues to fascinate millions of people today. Color section.
420 Pages. 6x9 Paperback. Illustrated. $22.00 Code: AOG

REICH OF THE BLACK SUN
Nazi Secret Weapons & the Cold War Allied Legend
by Joseph P. Farrell
Why were the Allies worried about an atom bomb attack by the Germans in 1944? Why did the Soviets threaten to use poison gas against the Germans? Why did Hitler in 1945 insist that holding Prague could win the war for the Third Reich? Why did US General George Patton's Third Army race for the Skoda works at Pilsen in Czechoslovakia instead of Berlin? Why did the US Army not test the uranium atom bomb it dropped on Hiroshima? Why did the Luftwaffe fly a non-stop round trip mission to within twenty miles of New York City in 1944? *Reich of the Black Sun* takes the reader on a scientific-historical journey in order to answer these questions. Arguing that Nazi Germany actually won the race for the atom bomb in late 1944, 352 PAGES. 6x9 PAPERBACK. ILLUSTRATED. BIBLIOGRAPHY. $16.95.
CODE: ROBS

THE GIZA DEATH STAR
The Paleophysics of the Great Pyramid & the Military Complex at Giza
by Joseph P. Farrell
Was the Giza complex part of a military installation over 10,000 years ago? Chapters include: An Archaeology of Mass Destruction, Thoth and Theories; The Machine Hypothesis; Pythagoras, Plato, Planck, and the Pyramid; The Weapon Hypothesis; Encoded Harmonics of the Planck Units in the Great Pyramid; High Freqguency Direct Current "Impulse" Technology; The Grand Gallery and its Crystals: Gravito-acoustic Resonators; The Other Two Large Pyramids; the "Causeways," and the "Temples"; A Phase Conjugate Howitzer; Evidence of the Use of Weapons of Mass Destruction in Ancient Times; more.
290 PAGES. 6x9 PAPERBACK. ILLUSTRATED. $16.95. CODE: GDS

THE GIZA DEATH STAR DEPLOYED
The Physics & Engineering of the Great Pyramid
by Joseph P. Farrell
Farrell expands on his thesis that the Great Pyramid was a maser, designed as a weapon and eventually deployed—with disastrous results to the solar system. Includes: Exploding Planets: A Brief History of the Exoteric and Esoteric Investigations of the Great Pyramid; No Machines, Please!; The Stargate Conspiracy; The Scalar Weapons; Message or Machine?; A Tesla Analysis of the Putative Physics and Engineering of the Giza Death Star; Cohering the Zero Point, Vacuum Energy, Flux: Feedback Loops and Tetrahedral Physics; and more.
290 PAGES. 6x9 PAPERBACK. ILLUSTRATED. $16.95. CODE: GDSD

THE GIZA DEATH STAR DESTROYED
The Ancient War For Future Science
by Joseph P. Farrell
Farrell moves on to events of the final days of the Giza Death Star and its awesome power. These final events, eventually leading up to the destruction of this giant machine, are dissected one by one, leading us to the eventual abandonment of the Giza Military Complex—an event that hurled civilization back into the Stone Age. Chapters include: The Mars-Earth Connection; The Lost "Root Races" and the Moral Reasons for the Flood; The Destruction of Krypton: The Electrodynamic Solar System, Exploding Planets and Ancient Wars; Turning the Stream of the Flood: the Origin of Secret Societies and Esoteric Traditions; The Quest to Recover Ancient Mega-Technology; Non-Equilibrium Paleophysics; Monatomic Paleophysics; Frequencies, Vortices and Mass Particles; "Acoustic" Intensity of Fields; The Pyramid of Crystals; tons more.
292 pages. 6x9 paperback. Illustrated. $16.95. Code: GDES

SAUCERS, SWASTIKAS AND PSYOPS
A History of a Breakaway Civilization
By Joseph P. Farrell

Farrell discusses SS Commando Otto Skorzeny; George Adamski; the alleged Hannebu and Vril craft of the Third Reich; The Strange Case of Dr. Hermann Oberth; Nazis in the US and their connections to "UFO contactees"; The Memes—an idea or behavior spread from person to person within a culture— are Implants. Chapters include: The Nov. 20, 1952 Contact: The Memes are Implants; The Interplanetary Federation of Brotherhood; Adamski's Technological Descriptions and Another ET Message: The Danger of Weaponized Gravity; Adamski's Retro-Looking Saucers, and the Nazi Saucer Myth; Dr. Oberth's 1968 Statements on UFOs and Extraterrestrials; more.
272 Pages. 6x9 Paperback. Illustrated. $19.95. Code: SSPY

LBJ AND THE CONSPIRACY TO KILL KENNEDY
By Joseph P. Farrell

Farrell says that a coalescence of interests in the military industrial complex, the CIA, and Lyndon Baines Johnson's powerful and corrupt political machine in Texas led to the events culminating in the assassination of JFK. Chapters include: Oswald, the FBI, and the CIA: Hoover's Concern of a Second Oswald; Oswald and the Anti-Castro Cubans; The Mafia; Hoover, Johnson, and the Mob; The FBI, the Secret Service, Hoover, and Johnson; The CIA and "Murder Incorporated"; Ruby's Bizarre Behavior; The French Connection and Permindex; Big Oil; The Dead Witnesses: Guy Bannister, Jr., Mary Pinchot Meyer, Rose Cheramie, Dorothy Killgallen, Congressman Hale Boggs; LBJ and the Planning of the Texas Trip; LBJ: A Study in Character, Connections, and Cabals; LBJ and the Aftermath: Accessory After the Fact; The Requirements of Coups D'État; more.
342 Pages. 6x9 Paperback. $19.95 Code: LCKK

THE TESLA PAPERS
Nikola Tesla on Free Energy & Wireless Transmission of Power
by Nikola Tesla, edited by David Hatcher Childress

David Hatcher Childress takes us into the incredible world of Nikola Tesla and his amazing inventions. Tesla's fantastic vision of the future, including wireless power, anti-gravity, free energy and highly advanced solar power. Also included are some of the papers, patents and material collected on Tesla at the Colorado Springs Tesla Symposiums, including papers on: •The Secret History of Wireless Transmission •Tesla and the Magnifying Transmitter •Design and Construction of a Half-Wave Tesla Coil •Electrostatics: A Key to Free Energy •Progress in Zero-Point Energy Research •Electromagnetic Energy from Antennas to Atoms
325 PAGES. 8x10 PAPERBACK. ILLUSTRATED. $16.95. CODE: TTP

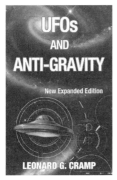

UFOS AND ANTI-GRAVITY
Piece For A Jig-Saw
by Leonard G. Cramp

Cramp first introduces the idea of 'anti-gravity' and introduces us to the various theories of gravitation. He then examines the technology necessary to build a flying saucer and examines in great detail the technical aspects of such a craft. Cramp's book is a wealth of material and diagrams on flying saucers, anti-gravity, suppressed technology, G-fields and UFOs. Chapters include Crossroads of Aerodymanics, Aerodynamic Saucers, Limitations of Rocketry, Gravitation and the Ether, Gravitational Spaceships, G-Field Lift Effects, The Bi-Field Theory, VTOL and Hovercraft, Analysis of UFO photos, more.
388 PAGES. 6x9 PAPERBACK. ILLUSTRATED. $19.95. CODE: UAG

ORDER FORM

10% Discount When You Order 3 or More Items!

One Adventure Place
P.O. Box 74
Kempton, Illinois 60946
United States of America
Tel.: 815-253-6390 • Fax: 815-253-6300
Email: auphq@frontiernet.net
http://www.adventuresunlimitedpress.com

ORDERING INSTRUCTIONS

✓ Remit by USD$ Check, Money Order or Credit Card
✓ Visa, Master Card, Discover & AmEx Accepted
✓ Paypal Payments Can Be Made To:
   info@wexclub.com
✓ Prices May Change Without Notice
✓ 10% Discount for 3 or More Items

SHIPPING CHARGES

United States
✓ Postal Book Rate { $4.50 First Item / 50¢ Each Additional Item
✓ POSTAL BOOK RATE Cannot Be Tracked!
   Not responsible for non-delivery.
✓ Priority Mail { $6.00 First Item / $2.00 Each Additional Item
✓ UPS { $7.00 First Item / $1.50 Each Additional Item
NOTE: UPS Delivery Available to Mainland USA Only

Canada
✓ Postal Air Mail { $15.00 First Item / $3.00 Each Additional Item
✓ Personal Checks or Bank Drafts MUST BE US$ and Drawn on a US Bank
✓ Canadian Postal Money Orders OK
✓ Payment MUST BE US$

All Other Countries
✓ Sorry, No Surface Delivery!
✓ Postal Air Mail { $19.00 First Item / $7.00 Each Additional Item
✓ Checks and Money Orders MUST BE US$ and Drawn on a US Bank or branch.
✓ Paypal Payments Can Be Made in US$ To:
   info@wexclub.com

SPECIAL NOTES

✓ RETAILERS: Standard Discounts Available
✓ BACKORDERS: We Backorder all Out-of-Stock Items Unless Otherwise Requested
✓ PRO FORMA INVOICES: Available on Request
✓ DVD Return Policy: Replace defective DVDs only
ORDER ONLINE AT: www.adventuresunlimitedpress.com

10% Discount When You Order 3 or More Items!

Please check: ✓

☐ This is my first order   ☐ I have ordered before

Name
Address
City
State/Province | Postal Code
Country
Phone: Day | Evening
Fax | Email

Item Code	Item Description	Qty	Total

Subtotal ▶
Please check: ✓   Less Discount-10% for 3 or more items ▶

☐ Postal-Surface   Balance ▶
☐ Postal-Air Mail   Illinois Residents 6.25% Sales Tax ▶
   (Priority in USA)   Previous Credit ▶
☐ UPS   Shipping ▶
   (Mainland USA only) Total (check/MO in USD$ only) ▶
☐ Visa/MasterCard/Discover/American Express

Card Number:
Expiration Date: Security Code:

✓ SEND A CATALOG TO A FRIEND: